## Praise for *Threat Multiplier*

"Climate change adds additional instability to a world already disrupted by strategic competition with China, Russia, and Iran, two regional wars, and mounting challenges from new technologies such as AI and cyber. Sherri Goodman shows how the US military is managing these risks and why a bipartisan approach to building a more resilient future supports our national security."

—Stephen J. Hadley, former US National Security Advisor

"As Secretary of Defense, my job was to protect the nation from all manner of threats. Climate change was one of those threats. But it's not just one among a list of many. It's the threat that multiplies all others. Sherri Goodman gives us a first-hand account of how climate change affects the military and our national security—and the steps we must take to build a more secure future."

—Leon Panetta, former Secretary of Defense

"When I led US forces in Europe and Latin America, I witnessed how a changing climate creates instability, from weather extremes to environmental degradation. *Threat Multiplier* is a must-read for military, civilian, and humanitarian leaders who want to understand how a changing climate and the energy transition affect decision making on these critical global challenges."

—General Wesley Clark, former NATO Supreme Allied Commander

"When the scientific understanding of climate change and its impacts was just beginning to solidify, Sherri Goodman realized that climate change would be at the very heart of military security, for the United

States and every other country. Her leadership on climate change as a threat multiplier has been the stimulus for a rich and dynamic body of scholarship and policy. This book connects the threads that link the science and the strategy. And it points the way to a safe world."

—Christopher Field, Director, Stanford Woods Institute for the Environment

"As Secretary of Defense, I observed how our allies and partners face direct risks to their stability from a changing climate. Sherri Goodman's insider's view shows us how our nation's military leaders are making decisions that improve our global resilience to climate change."

—Chuck Hagel, former Secretary of Defense and US Senator

"Sherri Goodman draws on her own experience and that of many military leaders to make a compelling case that climate change serves as a catalyst for conflict in vulnerable parts of the world and can weaken our security at home if we don't act now to build resilience to these complex geopolitical risks. Her message is a bipartisan call to act today to protect our national security in the future."

—Michael Chertoff, former Secretary of Homeland Security

"For anyone interested in climate change and US national security, *Threat Multiplier* is a must-read book. Goodman offers a compelling history of the Pentagon's pivot to consider climate change as a major security threat and provides invaluable insights to guide future defense policy in dealing with this growing danger."

—Michèle Flournoy, former Undersecretary of Defense for Policy

"When I led US forces in the Middle East, conflict, not climate change, was at the front of my mind. Yet, I could see how the two are inextricably intertwined, with climate change multiplying the threats that lead to conflict. Sherri Goodman makes those connections in this excellent book—one I would recommend to all military leaders."

—General Anthony Zinni, former Commander, US Central Command

"No one understands better than Sherri Goodman how a changing climate is reshaping our national security, and how leaders, both military and civilian, must integrate these risks and opportunities into our global future. *Threat Multiplier* offers an insider's perspective on how climate and energy affect national security strategy and our military operations."

—Fred Kempe, President and CEO, Atlantic Council

"As a conservation leader, I want to protect our planet's precious natural resources for future generations. Sherri Goodman shows us how the military is doing its part and why natural resource conservation and climate change are core components of global security, as seen through the eyes of the nation's top military leaders."

—Carter Roberts, President and CEO, World Wildlife Fund US

"Sherri Goodman tells the stories of how military leaders like me experience the changing climate in our missions, from drought to storms. She helps us understand how we can better prepare and adapt, showing that the threat multiplier of climate change can become the opportunity multiplier of smart solutions."

—General (Ret) Tom Middendorp, Chairman, International Military Council on Climate & Security, and former Dutch Chief of Defense

# Threat Multiplier

## CLIMATE, MILITARY LEADERSHIP, AND THE FIGHT FOR GLOBAL SECURITY

Sherri Goodman

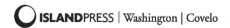

© 2024 Sherri Wasserman Goodman

All rights reserved under International and Pan-American Copyright Conventions. No part of this book may be reproduced in any form or by any means without permission in writing from the publisher: Island Press, 2000 M Street, NW, Suite 480-B, Washington, DC 20036-3319.

Library of Congress Control Number: 2024931902

All Island Press books are printed on environmentally responsible materials.

Manufactured in the United States of America
10 9 8 7 6 5 4 3 2

*Keywords:* Arctic warming; China illegal fishing; climate change; climate security; climate-vulnerable countries; CNA Military Advisory Board; Department of Defense; deputy undersecretary of defense (environmental security); environmental security; International Military Council on Climate and Security; military bases; military energy efficiency; natural disasters; nuclear disarmament; Pentagon; Russian oil industry; sea level rise; search and rescue mission; Senate Armed Services Committee

*To my parents and grandparents, whose decision to leave Nazi Germany made everything possible.*

*"I look forward to an America which will not be afraid of grace and beauty, which will protect the beauty of our natural environment . . . an America which commands respect throughout the world not only for its strength but for its civilization as well."*

—John F. Kennedy, remarks at Amherst College, October 26, 1963

# Contents

| | |
|---|---|
| Introduction | 1 |
| Chapter 1 From Weapons to Waste | 7 |
| Chapter 2 The Birth of Environmental Security | 21 |
| Chapter 3 Generals and Admirals Battle Climate Change | 43 |
| Chapter 4 Melting Ice and Rising Tensions in the Arctic | 67 |
| Chapter 5 Drought, Oil, and Power in Africa and the Middle East | 85 |
| Chapter 6 Navigating Asia's Disaster Alley | 103 |
| Chapter 7 Imperiled Neighbors to the South | 119 |
| Chapter 8 Climate Readiness on the Home Base | 139 |
| Chapter 9 Less Fuel, More Fight | 165 |
| Chapter 10 Climate-Proofing Security | 185 |
| *Acknowledgments* | *203* |
| *Notes* | *209* |
| *Index* | *239* |
| *About the Author* | *249* |

# Introduction

> *"Today, no nation can find lasting security without addressing the climate crisis. We face all kinds of threats in our line of work, but few of them truly deserve to be called existential. The climate crisis does."*
> —US Secretary of Defense Lloyd J. Austin

WHILE ATTENDING A LEADERSHIP SUMMIT on climate change in 2021, US Secretary of Defense Lloyd J. Austin called the climate crisis an existential threat, one that no nation can ignore if it wants to achieve lasting security. Roughly thirty years earlier, when I became the first deputy undersecretary of defense (environmental security) in US history, you would have been hard-pressed to find a military leader who worried as much about the stability of the planet as about the aggressions of our enemies. Indeed, through much of the twentieth century, the environment was at best a secondary consideration for much of the military. Our leaders were, after all, occupied with other matters, from deterring a nuclear attack by the Soviet Union to confronting terrorism. As a major global power, America needs the most capable, best-equipped, and best-trained fighting force in the world. Few could argue with our nation's military prowess, yet that effectiveness often came at a high cost to the planet. From dropping Agent Orange on the battlefield to dumping toxic chemicals on our soil, actions by the US military have had long-lasting health and environmental consequences at home and abroad. And yet, today, environmental stewardship is integral to the US military's activities.

This stunning turn has enabled the military to keep pace with a changing world. Our armed forces are now working actively to understand and reduce the risks from environmental degradation and a changing climate. And beyond our borders, militaries around the world have become climate leaders, establishing ambitious emissions goals, building installations that can stand up to the extreme weather of a warming world, and designing revolutionary clean energy technologies. At a moment when global climate leadership is desperately needed, our military has built on its past and stepped up to become a guiding light toward a more sustainable future.

What happened in the thinking of America's civilian leaders and military officers? Did the nation's generals go to bed one night as warfighters and wake up the next morning as environmentalists? Hardly. Instead, the idea of environmental security—the notion that our safety depends on a stable, healthy planet—is now deeply connected with concepts of mission and military readiness, across the administrations of both parties. This transformation occurred slowly at first but has accelerated as environmental challenges have increased. And it is still evolving today.

In some ways, military leaders have always been acutely aware that the state of the natural environment affects security. For generals to launch a ground offensive, they must know every detail of the terrain. Sending naval ships into battle similarly requires a sophisticated understanding of oceanography. Likewise, accurate weather forecasting has played a decisive role in many of history's pivotal battles, including the famous D-Day landings that helped put an end to World War II.

But for too long, environmental protection remained a blind spot for our military. How could defense officials be expected to worry about a few chemicals or the fate of polar bears when they were charged with saving human lives and defeating adversaries? The prevailing view among the uniformed military when I arrived at the US Senate Committee on Armed Services in the 1980s was that environmentalists simply did not understand the threats facing our nation or what was required to keep us safe.

Yet even in those early days, cracks began to form in the seemingly impenetrable wall constructed between national security and environmental protection. As the Cold War came to a close and the nation shuttered its nuclear bomb–making plants, the US Department of Defense (DOD) was forced to deal with decades of contamination—and, as I've often said, my career went from weapons to waste. Similarly, as groundbreaking environmental laws and regulations were enacted, including protections for endangered species on military bases, DOD had to figure out how to conserve natural resources while training the world's most elite fighters.

Slowly but surely, if not uniformly, attitudes began to change. The mere fact that in the 1990s, I was appointed to serve as the first deputy undersecretary of defense (environmental security) demonstrated that both the military leaders and the civilian officials they answered to were beginning to recognize that military readiness and environmental protection could go hand in hand. We cleaned up military bases, protected species' vital habitats, and shared these practices with nations around the world. Of course, there were naysayers: soldiers, sailors, generals, and plenty of civilians who thought environmental protection was mission creep that distracted our forces from their core job of defending the nation. But the progress was undeniable. The military's new awareness was not only good for trees and turtles; it was also good for the health and safety of our troops and the communities they served.

Then, in the early 2000s, we started confronting a new foe, one unlike any the world had ever seen, one that had been gathering strength for decades: climate change. From extreme storms to rising seas to drought and wildfire—and the political conflict and violence fueled by these stressors—climate change is a major international security threat. In fact, as I put it in 2007, it is a "threat multiplier," making every fragile state more vulnerable and every conflict more dangerous.[1]

In many ways, the military's environmental awakening in the 1980s and 1990s was excellent training to confront a warming planet. Our leaders had always relied on scientific evidence, and as health and environmental conditions changed, they adjusted military operations and training structures in response.

However, when the nation's generals first learned about climate change, it was considered a theory, one that was hotly contested in the media, if not in the scientific community. Moreover, this threat went far beyond any environmental challenge the military had faced before. It wasn't a matter of replacing harmful chemicals with benign alternatives or working around a sensitive habitat. Climate change would fundamentally shift the world's geopolitics, demanding new diplomatic strategies, new energy strategies for powering the force, and new approaches for protecting bases from natural disasters.

Today, we face the global threat of a rapidly changing climate that is reshaping our security environment. How do military leaders view this threat? How has their thinking evolved? What are their contingency plans? Where do they see the coming conflicts?

I can answer these questions because I've been on the front line of this fight for my entire professional life. I've debated generals in budget battles, asking that a small portion of our massive defense spending be used to address environmental threats. I've negotiated conflicts between environmental activists, who, understandably, want safe, healthy communities, and our armed forces, who, equally understandably, do not want to compromise military readiness.

In the 2000s, I convened a group of venerated military leaders to examine climate change as a threat multiplier. Their voices would cut through the political noise and show the world that an unstable climate puts global security and our national defense in serious peril. Indeed, the field of climate security sprang from the initial work of these leaders, known as the CNA Military Advisory Board, and then sprouted into a thriving international community of practice led by the Center for Climate and Security, the Climate and Security Advisory Group, and, later, the International Military Council on Climate and Security.

In the following pages, I share their stories and my own. We will travel to the top of the world, where collapsing permafrost and retreating sea ice in the Arctic are transforming trade and intersecting with growing competition among Russia, China, and other nations. We will follow military leaders to Africa and the Middle East, where they witnessed how violent extremist groups are weaponizing water and manipulating vulnerable communities who are already hit hard by climate change. In the Indo-Pacific region, we will see how the military is working to counter China's growing influence while actively supporting allies whose very sovereignty is threatened by sea level rise and a lack of fresh water. In Latin America, we will examine our own role in the mining of critical minerals—pursuits that can undermine already fragile states, whose citizens are migrating north. And on the home front, we will track the military's efforts to cut our tether to fossil fuels and prepare our bases for the raging storms to come.

I wrote this book to help readers understand the severe security threats that climate change presents and the hard choices that our military must make to defend against them. By putting you in the room and on the battlefield with generals and other military leaders, I hope to convey not only the complexity and danger of these risks but also the opportunity that the military is seizing. With the right strategies, from transforming the energy systems that power our tanks and aircraft to partnering with allies to improve weather forecasting, we can prepare our forces and our families for the most existential threat of our time.

It is not a straightforward task; there will continue to be doubt and disagreement about how best to protect ourselves in a dangerous world, where climate is one of many compounding factors. Yet it would be a dereliction of duty to turn away from this fight—and the military leaders with whom I have spent my career have no intention of doing so. They are already making tremendous progress, and there is still so much more to do. I make recommendations in the final chapter about how to further integrate climate into national security, from improving predictive capability to transforming energy use for a warming world.

Finally, I want to celebrate the individuals who have put their careers and their lives on the line to defend our nation in a warming world: The young officer in Afghanistan who inspects fuel trucks entering his base, knowing full well that the next one could contain explosives. The US Navy commander who leads her ships to provide aid in the aftermath of Superstorm Sandy. The admiral who is willing to tell a skeptical US Congress that climate change is among the most serious dangers he faces.

Experiences such as these have taught our armed forces the imperative of resilience in the face of an unprecedented global threat. Against all odds, and in defiance of all stereotypes, the military has become one of the world's strongest forces in the fight against climate change, and not solely for the sake of the climate. The military's focus is about ensuring it can continue its primary mission: to protect and defend the nation. In today's multipolar world, with US forces engaged around the globe, protecting America and gaining the upper hand against adversaries means being climate-smart.

To understand how we got here, we need to begin with the Cold War, as we will do in the next chapter.

CHAPTER 1

# *From Weapons to Waste*

Arriving for my first day at the Russell Senate Office Building was like entering a museum as a child. The space was grand, with large marble columns and high ceilings, but the furnishings were bureaucratic government-issued: deep fake-leather couches and old wooden desks with mismatched chairs. A female receptionist politely ushered me in, but her stare seemed to confirm my fear that I didn't belong here. I was shown to an empty desk in a musty back room and told I could work there until the whole office was renovated months later.

It was September 1987, and I had just been hired as the youngest and one of the first female professional staff member of the US Senate Committee on Armed Services, chaired by the Senate's authority on defense, Senator Sam Nunn of Georgia. Walking into my first staff meeting, I really knew I was out of place. When the staff director, Arnold Punaro, asked me to introduce myself, the one person I had already met piped up that he was pleased to introduce "Sherri Wasserman" to the committee. Punaro said, "Well, Sherri, aren't you married now?" I nodded: indeed, my wedding had been just the month before. Punaro replied, "Well, then your name is Sherri Goodman." I was so flummoxed that I couldn't utter a word. It wouldn't have mattered if I had managed to explain that my legal name was now Sherri Wasserman Goodman. The staff director's pronouncement was sufficient: as a married woman, I would have my husband's name. It was as simple as that.

I was an unlikely hire for Punaro and the Senate Armed Services Committee in more ways than one. Punaro was a retired marine, with portraits of Confederate generals hanging in his office. I was the daughter of Holocaust refugees, raised in New York and educated at colleges and universities in New England. In some ways, I had gotten used to being one of the few women in the room. In 1977, I had joined the second class of women to enter the formerly all-male bastion of Amherst College, where I fell in love with diplomatic history and international affairs. It was during a year at the London School of Economics and Political Science that I met several officials from the US Department of Defense (DOD) at a conference: encounters that would send me down a path that ultimately led to room 222 at the Russell Senate Office Building. I thought I was launching my career in nuclear weapons and arms control; little did I know that this would become a gateway to pioneering the fields of environmental security and, later, climate security.

I was fortunate to arrive at the Senate Armed Services Committee armed with good degrees and powerful professional connections. (At Harvard Law School, I had been classmates with Elena Kagan, now a justice of the US Supreme Court.) My mentors had unparalleled national security bona fides, and they were willing to ask hard questions and level criticism when it was due. When I was in college, a deputy undersecretary of defense for policy, Walter Slocombe, suggested I write my thesis on President Jimmy Carter's failed effort to deploy the neutron bomb, known as the weapon "that kills people and leaves buildings standing," into the arsenal of the North Atlantic Treaty Organization (NATO). That became my first book, *The Neutron Bomb Controversy:*

*A Study in Alliance Politics*.¹ Later, at Harvard, I studied with a former undersecretary of the navy, Robert J. Murray, who was leading a project that examined the "waste, fraud, and abuse" in defense contracting that became notorious during President Ronald Reagan's military buildup in the mid-1980s. That research led me to a critical examination of DOD's contracting practices and my first law review publication.² I learned during those years that you don't need to be an activist to create change; you can challenge the status quo by well-reasoned analysis from within the walls of the establishment.

Perhaps it was this willingness to think critically that Arnold Punaro responded to during my interview with him. He later told me that, while he usually never hired anyone without fifteen years of experience, I had impressed him. For Senator Nunn, my unusual perspective was an asset rather than a liability: there was a "gap in oversight" of the US Department of Energy's weapons complex, and they needed someone who would "challenge the basic assumptions" and not just "come in and salute."³ He told me, "We wanted someone with a fresh pair of eyes and a big brain."⁴ This turned out to have been the real reason Senator Nunn and Punaro tasked me with overseeing the nation's nuclear weapons plants. For many years, I've been half-joking that I got that assignment because no one else wanted it, but Punaro told me I shouldn't sell myself short. Looking back now, I can see that the man who identified me by my husband's name called out a trait I often observe in young women but had failed to recognize in myself. I doubted my own abilities, even when others praised them.

In this case, the praise came from an authoritative source. Senator Nunn had ascended to chair of the Senate Armed Services Committee after the 1986 midterm elections, when Democrats recaptured the Senate six years into Ronald Reagan's presidency. At forty-nine, Nunn was already legendary in defense circles. Elected to the Senate in 1972, he was known for his deep knowledge of and affection for America's

Figure 1-1. Senator Sam Nunn, Robert Murray, and Sherri Goodman, Pentagon, 1996 (photo credit: author's personal collection)

military. Nunn traced his political lineage to the congressional defense barons Senator Richard Russell and Congressman Carl Vinson, both of Georgia, who had ruled the roost as chairs of the Senate Armed Service Committee and the US House Committee on Armed Services. Indeed, Senator Nunn was a grandnephew of Congressman Vinson and the latest in a long line of defense-oriented Southern Democrats, whose chief power and purpose was protecting our troops and ensuring they were trained and equipped to fight America's wars.

Senator Nunn, however, prided himself on not being a rubber stamp for either the military or the president. The Senate Armed Services Committee largely exercises its power through the annual National Defense

Authorization Act, which directs policy and programs for defense spending and is one of the few major legislative vehicles passed each year by Congress. In this role, Senator Nunn wanted to be recognized for his independence and integrity. And he had plenty of both. He worked hard to learn the defense budget and to show the military leaders that when he held them to account, he did so with their best interests in mind.

Senator Nunn's formative experiences had taken place in the shadow of the Cold War's major nuclear crises, particularly the Cuban Missile Crisis, during which, as a twenty-four-year-old congressional staffer, he found himself on a delegation at NATO. Those early days gave him a visceral knowledge of the dangers of nuclear weapons, and he devoted most of his life to preventing the horrors they could inflict. I was now that young staffer, witnessing new threats posed by the nuclear age and charged with keeping them in check.

Since World War II, the United States had built dozens of nuclear reactors and related processing plants to make the fissile materials needed for nuclear warheads. These weapons were, after all, the centerpiece of the Cold War with the Soviet Union. As the only superpowers with nuclear capabilities, the two nations were locked in a precarious pact of mutual assured destruction (MAD): a deliberate balance of nuclear weapons that was supported by facilities scattered across remote parts of the United States and the Soviet Union.

Now I was responsible for overseeing the billion-dollar budgets that the Senate Armed Services Committee would annually authorize to fund these plants. And that meant one of my first assignments was to visit them, often in the middle of nowhere. In the high desert of eastern Washington, I toured the reactors cranking out fissile materials for nuclear weapons and met the engineers and other workers who were putting their lives and health on the line every day. I visited the Rocky Flats plutonium-processing plant outside Denver, where plutonium pits were fabricated, and was warned not to enter the infamous "glovebox" building, where plutonium parts were assembled, because I was of "childbearing age."

If it was too dangerous for me to make even a short visit, what did that mean for the workers inside, slipping their hands into extended rubber gloves through a glass wall to manipulate plutonium parts for nuclear warheads? Within months, the Federal Bureau of Investigation was conducting overflights of Rocky Flats, leading to criminal charges against the plant for safety and health violations. Pit production halted not long after I visited the plant, in 1989, and it never reopened.

**Reimagining Safety**

At first, it seemed that my job was to oversee a functioning set of facilities churning out the raw materials for weapons needed to ensure Americans were safe from the Soviet nuclear threat. But the more I learned, the more questions I had. What if that "safety" also endangered the lives of the people working in the plants and those beyond its walls? I was hardly the only one with these concerns. And they were put in stark relief by improvements made in civilian nuclear power (but not defense) plants after the partial meltdown at the Three Mile Island nuclear facility.

The accident was the worst in the history of the US nuclear power industry.[5] It started at 4:00 a.m. on March 28, 1979, when a pressure relief valve in the Unit 2 Reactor at Three Mile Island, Pennsylvania, failed to close. While no one died as a direct result of the accident, two million people were exposed to small amounts of radiation and 140,000 people evacuated the area.[6] After that partial meltdown—the only nuclear disaster on US soil—civilian nuclear power plants underwent substantial reforms. The US Nuclear Regulatory Commission increased its safety oversight of civilian reactors to ensure such an incident could never happen again.[7]

These reforms, however, did not extend to the nuclear weapons complex. Defense reactors produced plutonium and tritium for nuclear bombs, not power for homes and businesses. Their operators answered to a less rigorous safety authority at the US Atomic Energy Commission and, later, the US Department of Energy (DOE), whose priority was producing warheads in the arms race with the Soviet Union. Indeed, the nuclear weapons facilities operated as if many of the nation's environmental laws did not apply. Local opposition to these facilities had been building for years, fueled by concerns about polluted water and sloppy safety practices. Then, in 1986, the Chernobyl nuclear disaster brought further scrutiny of US defense reactors.[8] If a meltdown could happen outside Kiev, could it also happen at the Savannah River Site in South Carolina or on the banks of the Columbia River in Hanford, Washington?[9]

By the fall of 1988, the *New York Times* was running front-page stories almost every week about safety lapses at the nation's nuclear weapons plants.[10] From the first of October to the end of December, eighty-five stories on the issue appeared in the paper, with headlines such as "Accidents at a U.S. Nuclear Plant Were Kept Secret Up to 31 Years" and "Chronic Failures at Atomic Plant Disclosed by U.S. High Rate of Shutdowns."[11] As one *Times* reporter recounted, "Our editors haven't been this psyched about a story since the Pentagon Papers."[12]

With hard copies of newspapers appearing on their desks every morning, members of Congress started paying attention. It was my job to report each new concern at nuclear weapons plants to the Senate Armed Services Committee, including to my direct boss, Senator Nunn, whose Savannah River Site, where many Georgians worked, was among those chronicled. With each successive story, the senators asked more questions. The committee held hearings on the issue, and my career went "from weapons to waste." Senator Nunn directed me to draft legislation creating a new oversight mechanism for these nuclear weapons plants. I got my first lesson in legislative politics when I went up against another committee that also claimed jurisdiction on this bill—the formidable US Senate Committee on Governmental Affairs.

Senator Nunn knew that it was time for better oversight of the nuclear weapons complex, and having the Armed Services Committee write the legislation would both preserve the committee's prerogatives and ensure that the committee that funded these facilities stayed in control. However, another member of the Armed Services Committee and the chair of the Senate Governmental Affairs Committee, Senator John Glenn of Ohio, and his senior staff had other ideas. One of the National Aeronautics and Space Administration (NASA)'s original seven Mercury astronauts, Glenn was considered a national hero.[13] His home state also had nuclear weapons facilities, including a uranium-processing facility in Fernald, where local citizens were already raising alarms about the safety of working at the plant. Senator Glenn believed strongly that the nuclear weapons plants should be governed as the civilian power plants were. And his staff director was eager to take on a battle with my staff director because they had clashed on other matters in the past.

The legislative battle lines were drawn: the Governmental Affairs Committee sought jurisdiction over the legislation that Senator Nunn had directed me to draft. That set me up against Steve Ryan, an experienced prosecutor then serving as general counsel to the Governmental Affairs Committee. Ryan would later become one of the go-to lawyers in Washington, DC, for congressional investigations, including by representing Michael Cohen, President Donald Trump's personal lawyer. At the time, Ryan reported to his staff director, Leonard Weiss, a nuclear engineer who had been the chief architect of the Nuclear Non-Proliferation Act of 1978. They were both pushing for as rigorous a regulation of the nuclear plants as they could convince a majority of the members to adopt. My marching orders were to give very little to their demands, regardless of how much they yelled at me, which was not infrequently. I often felt on the verge of being outwitted by more experienced staff, but eventually, after months of negotiating, we landed on a compromise: what is today called the Defense Nuclear Facilities Safety Board.[14]

Like many legislative battles, ours ended in a territory where no one was entirely happy. The safety board would be part of the executive branch and provide recommendations for the nuclear plants to the US president and the secretary of energy. But it would not provide for the formal "notice and comment" regulation that the Nuclear Regulatory Commission conducted of civilian nuclear power plants. That type of regulation would have moved oversight from the executive branch to an independent regulatory body, which neither the Armed Services Committee nor the Reagan administration wanted: the Armed Services Committee wanted to tighten the safety processes within DOE while retaining the committee's ability to conduct oversight and provide funding as its principal sources of leverage. As it turned out, with the Cold War waning and the Soviet Union collapsing, many of these plants, such as Rocky Flats, eventually closed down. DOE reorganized the closed facilities into an Environmental Management program, consisting of many Superfund sites, part of a vast cleanup program with separate management, leaving only a small set of plants still operating within the oversight of the new board.

Little did I know that I had arrived at the tail end of the nuclear weapons production era. In 1988, the secretary of energy declared that the United States was "awash in plutonium," meaning there was no need to produce more.[15] This statement, along with the growing safety and environmental risks, became the nail in the coffin for the nation's nuclear weapons complex. In my first few years on the committee staff, I even witnessed one of the last underground nuclear weapons tests in the country. Accompanying members of the Nuclear Weapons Council, I traveled to an expansive control room in the middle of the Nevada desert. As the explosion rocked the terrain several miles away, all we felt was a slight shake.

Yet the fight over regulation of the nuclear sites was a transformative experience. I learned the power that committees such as the Armed Services Committee had to shape the direction of vast agencies. A decision made in a single committee room could have profound implications for DOD and DOE, whose work went to the heart of America's national security. I also learned how power in the Senate is wielded. Passing major legislation in the Senate requires garnering support from many different quarters, and Senator Nunn and his staff director, Arnold Punaro, were master legislative tacticians. They knew that coalitions are not always created by party affiliation. When it came to regulating the nuclear weapons complex, for example, members did not vote red or blue but on the basis of whether they had nuclear plants in their districts.

Maybe most importantly, I learned that even a carefully constructed national defense strategy can be overturned by real-world events, such as a nuclear disaster. As the military saying goes, "No plan ever survived first contact with the enemy." In this case, the enemy was "safety" being defined too narrowly. Public opinion would not support a defense strategy that protected Americans from foreign powers but threatened their health and safety at home. I would watch this movie—the convergence of the environment and the military—many times during the course of my career.

## Innovating Environmental Technology for Defense

As Senator Nunn thought about the growing environmental problems at the nation's now mostly defunct nuclear weapons facilities, he returned to an idea that had motivated his own career: the talent and commitment of public servants across the defense enterprise, from our uniformed military to the scientists and engineers in DOD and nuclear weapons labs. Senator Nunn made a point of visiting our national labs on a regular basis to keep abreast of key scientific and technological developments by the people who were "smarter than him." The highly regarded Los Alamos National Laboratory in New Mexico, Lawrence Livermore National Laboratory in California, and Sandia National Laboratories in New Mexico and California were created as part of the Manhattan Project to develop nuclear weapons. Some of the nation's most talented scientists flocked to these labs during the Cold War in the way that scientists today want to work in the tech meccas of San Francisco, Seattle, and Boston. These defense labs, at both DOD and DOE, were the place to be for the best and brightest physicists and engineers in the nuclear age, such as Robert Oppenheimer.

Now that the Cold War was over, Senator Nunn reasoned, their talent could be put to equally good use in addressing the nation's environmental challenges. And Senator Nunn had an eager partner in this endeavor: Senator Al Gore of Tennessee. Al Gore was a young and ambitious senator in the 1980s, having already made a run for the presidency in 1988. He had carved out his congressional profile in technology and environmental leadership. After joining the US House of Representatives, Gore held the first congressional hearings on climate change and cosponsored hearings on toxic waste. In 1990, he presided over a three-day conference with legislators from over forty countries, which sought to create a Global Marshall Plan, designed to help developing countries grow economically while still protecting the environment.[16]

Gore was also a member of the Senate Armed Services Committee and one of the early advocates on this committee of considering environmental matters at both DOD and DOE. Senator Nunn and Senator Gore joined forces to create the first Strategic Environmental Research and Development Program (SERDP) for the military. The SERDP Council is composed of officials from DOD, DOE, and the US Environmental Protection Agency (EPA); advisors from the President's Science Advisor and NOAA Administrator; and representatives from state government and nongovernmental organizations.

The program would, among other priorities, develop technologies for the toughest environmental problems the federal government faces, from radioactive and hazardous waste in the 1980s and 1990s to per- and polyfluoroalkyl substances (PFAS), the "forever chemicals" found at both military and civilian locations today. It has also developed technologies to remove unexploded ordnance from ranges, and it has trained almost half of the country's experts in wildland firefighting, a specialty that has become increasingly important as forest fires have grown into a national crisis.[17]

Little did Senators Nunn and Gore know when they created this program in 1990, with just $25 million in federal funding, that it would grow to lead the environmental technology revolution. They also did not know, nor did I, that this program would become the engine of environmental technology development so crucial to my work in the 1990s at DOD. There, I was in charge of cleaning up over one hundred bases listed as Superfund sites and needed to ensure that DOD was complying with the nation's major environmental laws.

**Unusual Partnerships**

While I was dealing with defunct nuclear plants, a young US Air Force officer was having his own environmental moment. In 1987, Tom Morehouse, who would later become my first military assistant, was an engineer at the Air Force Research Laboratory at Tyndall Air Force Base in Florida. He was tasked with improving chemical agents used to fight fires. Morehouse was sitting at his desk when an EPA official, Steve Anderson, walked up and informed him, "What you're doing is destroying the ozone layer."

Until that moment, Morehouse had not considered the environmental impact of his work. Anderson walked him through the science of how the chemical halon destroys ozone. "I was persuaded that there was something there, and it became obvious to me that the people who sell halon had done an extremely good job on selling it for purposes for which it was not required," Morehouse told me years later.[18]

The bulk of the air force's halon was used for training and testing rather than firefighting. Once Morehouse began looking at alternatives that were less damaging to the ozone layer, he found that the air force needed halon for only about 5 percent of the prior uses.[19] It also didn't need to flood entire rooms; the military could use halon in a more targeted way to manage fires. "Basically, our research showed that this was not a good story for halon, in terms of the size of the market as it existed at that time."[20]

Anderson was so impressed by the research that he asked Morehouse to present it at the negotiations on the Montreal Protocol on Substances that Deplete the Ozone Layer. It was the beginning of a friendship and professional partnership that has lasted for thirty-five years. Morehouse and Anderson have coauthored numerous peer-reviewed articles on alternatives to halons, and Morehouse's groundbreaking work helped lead the air force, and later DOD, to develop better substitutes for destructive chemicals.

What was unique about this collaboration? It marked one of the first times that military research was used to both improve defense capabilities and address a global environmental problem. Such partnerships would become increasingly crucial as environmental threats compounded, making ozone protection seem like a relative cakewalk. By the time I arrived at DOD six years later, in 1993, the military would be on the front lines of even more dire environmental challenges.

CHAPTER 2

# *The Birth of Environmental Security*

IN 1993, AT AGE THIRTY-FOUR, I was appointed deputy undersecretary of defense (environmental security). I was not only one of the youngest senior staff members at the Pentagon, not to mention one of the few women, but also the first person to hold the position. There is no denying that I take pride in that "first," yet I also recognize it was a battlefield promotion. In those days, not many people had experience with both military and environmental issues, and my work on the US Senate Committee on Armed Services was ideal training for the newly created role.

My early days at the Pentagon were almost more intimidating than the ones at the Senate six years earlier. Here I was at the heart of America's national security enterprise. Armed guards stood sentry at the entrance. Photos of the president, the secretary of defense, and the chairman of the Joint Chiefs of Staff stared out at visitors as they passed through security. Men and women, though mostly men, in dress uniforms scurried around the vast halls carrying important folders of decision memorandums called "packages."

My office was on the third floor of what is known as the E-ring, along the perimeter of the five-sided labyrinth. That location, just down the hall from the office of the secretary of defense, may have been chosen for a reason. While I wasn't the first person at the US Department of Defense (DOD) to oversee environmental programs, the new title of deputy undersecretary meant higher stature, more staff, and greater authority. I would have a seat at the most senior tables in defense decision-making. The new position sent a strong message about the incoming administration's commitment to sustainability. While Bill Clinton had centered his successful 1992 presidential campaign on economics, his choice of Al Gore as running mate showed he was also serious about environmental conservation. In addition to Gore's environmental work on the Senate Armed Services Committee, he had written a best-selling book, *Earth in the Balance*, and would become the environmental conscience of the administration. Both men were determined to make the environment a priority throughout the government—including in DOD.

While Clinton and Gore elevated environmental issues in the administration, they were not just Democratic issues. In the 1990s, protecting the planet had yet to become the partisan plank it is today. Indeed, during the 1992 campaign, former Republican president George H. W. Bush had made a point of speaking at the United Nations Conference on Environment and Development in Rio de Janeiro, Brazil (the Earth Summit), declaring it a historic gathering. The event galvanized international action on environmental and development issues and established the United Nations Framework Convention on Climate Change (UNFCCC), the foundational treaty for international climate negotiations. President Bush embraced the Earth Summit's critical mission, remarking, "The Chinese have a proverb: If a man cheats the Earth, the Earth will cheat man. The idea of sustaining the planet so that it may sustain us is as old as life itself. We must leave this Earth in better condition than we found it."[1]

His speech reflected the emerging international consensus that the state of the environment is a security issue. The end of the Cold War and the birth of the modern environmental movement set the stage for this new understanding. As I saw firsthand, the military no longer needed to build new nuclear weapons to fend off Russia, but it did need to deal with many decades of contamination from installations and past deployments. Yet the changes at DOD went further than cleaning up pollution. Growing environmental awareness and tougher enforcement of environmental laws also reshaped virtually every dimension of the military's activities, from training and education to the way we buy weapons systems to environmental considerations in overseas deployments.

It's important to remember that at the time, most media coverage of environmental issues in defense was extremely negative, with many health and environmental organizations criticizing lax military practices. "Military pollution at both DOD and DOE facilities will remain one of the nation's most critical problems for years to come," stated Physicians for Social Responsibility in a report issued just after I entered office.[2] While many of my DOD colleagues would have taken issue with that statement, they did see the need for a forward-looking strategy. As I stated in one of my first congressional testimonies, "The administration recognizes that the post–Cold War era requires a new approach to solving DOD's environmental problems. . . . Our national security can and must include protection of the environment."[3]

By the time I arrived at the Pentagon in early 1993, the battlefield had been prepped, so to speak. The outgoing administration had already elevated the environment to a global security concern, and the incoming Clinton-Gore administration wanted to deepen those efforts. At first, this ran headlong into DOD's traditional notion that environmental considerations were at best secondary to military activities and often were just a plain old nuisance; environmentalists, in this view, simply did not understand or appreciate the military and its mission. Fortunately,

I had some champions of environmental change at the highest levels of DOD, including the secretary and deputy secretary of defense and their senior advisors, Rudy de Leon and Larry Smith. Their advice was particularly helpful when I found myself in bureaucratic hot water.

## A Budget Battle for the Environment

I had been working at the Pentagon for less than six months when I attended my first budget meeting chaired by my boss, John Deutch, the hard-charging undersecretary of defense for acquisition, technology, and logistics. In the early 1990s, we often gathered in a windowless SCIF (sensitive compartmented information facility), not always because we were discussing classified information but because the undersecretary and his staff controlled attendance at this secure conference room. I was a young, small-statured woman in a room full of six-foot-tall, mostly White men with chests full of ribbons.

It didn't help that I was overseeing a newly elevated portfolio at the Pentagon—DOD's environmental programs—and many of my colleagues were skeptical. Very skeptical. My portion of the budget, admittedly, was a minuscule fraction of the money devoted to buying big weapons systems; supporting the world's largest research and development initiatives; and moving logistics, such as ammunition, food, and other supplies, to troops at the front. In other words, I didn't expect the discussion to focus much on environmental spending. Still, it was very important to me and the secretary of defense; it was the primary reason I was serving as the first deputy undersecretary of defense (environmental security).

Our initial focus was pollution. Congress had been putting pressure on DOD to clean up its mess since about one hundred military bases were listed on the Superfund registry of the most toxic sites nationwide. Citizens around those bases were becoming increasingly frustrated with the slow pace of progress. The Office of the Secretary of Defense, where

I worked, oversaw the cleanup program, allocating money to the US Army, US Navy, and US Air Force for specific projects. When I arrived at the Pentagon in 1993, funding for the program averaged about $1.5 billion per year, a significant sum, but small potatoes compared both with other aspects of military spending and with what would later be required.

Before my budget meeting, the environmental staff in the three military service branches had given me spreadsheets of data detailing the cost of cleaning up a wide variety of sites containing everything from perchlorinated solvents to leaking jet fuel. I reviewed this material and assembled a budget request for the next fiscal year, FY 1995, and prepared to defend it. I practiced my script. My slides, with the budget charts and justification, were ready.

Our meeting in the cavernous SCIF was also a test for my boss, who was running his first annual budget review. If he didn't win his budget battles, the military services would continue to push him and would not take his no for an answer. I listened carefully as each service budget chief stood up and presented his "requirements" for large weapons systems. Every man (and they were all men) had decades of experience and a commanding presence. In his green uniform, an army general presented his budget for tanks and other ground vehicles. In a sharp blue suit bedecked with ribbons, the air force general laid out his spending for fighters, bombers, and space-based systems. In a polished black jacket, the navy budget chief detailed his requirements for shipbuilding and naval aviation. My boss joked with each of them as if they were old buddies.

My turn came close to the end. I stood up, with none of the swagger of my predecessors. I gingerly stepped to the front of the room. My slides appeared on the screen. I carefully explained the budget request for DOD's environmental cleanup program, based on the numbers that each of the military services had provided. One by one, each of the service budget chiefs scoffed at me and my slides, saying they had never seen these figures before. "This isn't in my budget." "I haven't validated these numbers." They were playing a classic bureaucratic game I had yet to learn: if you don't accept the validity of a proposed budget, you may not have to pay for it. The budget chiefs knew that the money for base cleanups had to come from somewhere; it just wasn't going to come from their pockets. I couldn't believe it. My environmental counterparts in the services had set me up for failure.

My boss looked at me for a way out. I explained that the staff in the military services had given me these figures. But the budget chiefs were not buying it, literally. With bigger fish to fry, my boss chose not to support my proposed budget. I was left with hat in hand and feeling very embarrassed.

The navy's budget chief, known as the N-8, Vice Admiral Joe Lopez, who watched me go down in flames at this meeting, consoled me later with several important pieces of advice. First, "A strategy without validated funding is hallucination." You can have a lot of good ideas, but if you can't command the funding to put them into action, those ideas will die. In an interview for this book, Lopez conceded to me, "I wasn't going to give up a dime." The military was worried about budget cuts to weapons systems after the Cold War, and Lopez was feeling "strapped." "I was only interested in ships and planes." Reflecting back after almost thirty years, he mused: "God bless you for laying this [environmental budget] on the table, because I think all of us are a little parochial."[4]

A second important lesson I learned is that program offices—such as the environmental staffs of the army, navy, and air force—are not always connected to the powerful budgeteers who manage defense spending. Indeed, staff members of these offices weren't even present at the budget meeting and apparently had no idea what trade-offs the service chiefs and secretary of defense were making across the entire budget.

I was determined to never be caught flat-footed again. After that budget meeting, I set out to gain support from key DOD leaders. I knew

I'd need their trust if I hoped to make a difference. In the arcane battles of the defense budget, where money is power, this was perhaps my most important lesson. It was good to learn it early, even if I learned it the hard way.

## The Politics of Cleanup

As it turned out, my initial budget battle was part of a much larger war of competing economic interests. In the case of closing military bases that were no longer needed, DOD and environmentalists found common ground. But that didn't make the politics of the closures any easier.

The military leaders had wanted to close excess facilities for years because it would help their budgets, but congressional politics had always stood in the way. What elected official would vote to close a base in their district when it meant getting rid of well-paid jobs for their constituents? After many years of failure, Congress, under the leadership of my former boss, Senator Sam Nunn, set up an independent commission to select which bases to close. The US president couldn't cherry-pick but instead had to accept or reject the entire list. Several rounds of closures had already occurred, in 1988 and 1991, with communities complaining about economic hardship.

Scores of military bases were on the 1993 list, including many in California. That meant job losses in a state with a lot of voting power—and political backlash. Senators Dianne Feinstein and Barbara Boxer, who helped deliver California for Clinton with their own election in 1992, were not happy. Nor was the rest of the California congressional delegation, most of whom had worked hard to get Clinton elected. At the White House, the hand-wringing was so loud it reverberated across the Potomac to the Pentagon.

Although Clinton considered rejecting the list entirely, that decision would have further soured his relations with the military, especially because closing bases was economically necessary. The relationship had already been strained when, early in his administration, Clinton announced he would allow gay people to serve openly in the US military. He did not anticipate the backlash among many in the military and their supporters in Congress. The ultimate compromise, Clinton's infamous "Don't Ask, Don't Tell" policy, prohibited service members from disclosing their sexual orientation in exchange for superior officers not terminating their military service.[5] It was awkward and uncomfortable. At the Pentagon, it was widely thought that the president had just lost a round to the military.

This episode cost President Clinton valuable capital with the military that he could scarcely afford to squander. With the Cold War over, defense budgets were already on the decline, and now the uniformed military had another reason not to trust these young staffers in the White House. Moreover, the new secretary of defense, Les Aspin, was a policy wonk who preferred to spend his time on the strategic issues of denuclearization of the Soviet Union and global engagement. Unlike Senator Nunn, whose instincts were a good barometer of the uniformed military in that era, Aspin was more comfortable with the Ivy League professors he invited to serve as his lieutenants in the policy office of the secretary of defense.

When it came to base closures, the president knew he had little choice. Excess military facilities were reducing funding that might better be invested in the fighting tooth (training troops and buying weapons) of the armed forces. He would have to accept the list of closures and try to soften the blow to affected communities. The White House's plan involved developing a series of grants and economic programs—and I was tasked with spearheading DOD's portion of the strategy. Ultimately, we came up with measures that would support the communities' transition and save DOD about $12 billion over five years.[6] On July 2, 1993, Clinton announced his decision to close the military bases but flipped the script by emphasizing his "job-centered" closure plan.

To sell the plan, he flew to Naval Air Station Alameda near Oakland, California, to meet with community leaders and local officials.[7] I was part of the team traveling with the president, and my friend at the National Security Council, Nancy Soderberg, arranged to have me join Clinton for one of his famous public jogs. I got a call late in the day from an aide who told me where to meet the president and his security detail the next morning. In my purple running shorts and cotton T-shirt, I arrived at the appointed hour. The president greeted me with a grin and chatted with me about my road to DOD during our short ride in an armored limousine. After arriving at Lake Merritt, we started off, surrounded by muscled US Secret Service agents with earpieces. Fans and friends lined the lake, many shouting "Happy birthday, Mr. President!"—this was just a few days before August 19, when he would turn forty-seven. We were nearing the end of the loop when, without warning, President Clinton sprinted off, to applause from the crowd.

It wouldn't be the last time that day the president left me in the dust. At the public event later that morning, Clinton described his five-part plan to create jobs while closing bases. A key portion was speeding up the cleanup of contaminated lands to turn the property over to the community for development. The tension was that the studies, reviews, and remediation required by environmental laws would take many years to accomplish. When we prepared the draft of the president's plan at DOD, I recommended we call the cleanup "commonsense" to avoid false promises about the timeline. To the White House speechwriters, however, this phrase did not convey the necessary urgency, so they simply changed it to "fast-track."

The White House speechwriters, however, did not know how hard it had been just to broker an agreement between DOD officials and the Environmental Protection Agency (EPA) on what we had called "commonsense cleanup." Defense officials often thought environmental regulations created more red tape than actual cleanup. Meanwhile,

Figure 2-1. Jogging with President Bill Clinton, Oakland, California, August 14, 1993 (photo credit: author's personal collection)

many EPA officials didn't trust DOD to follow the rules necessary to protect public health. And EPA had the authority to review and approve DOD's cleanup work. So how was I, then only months into my tenure at the Pentagon, supposed to bring the warring factions together? That was where the art of multilevel negotiations came in. Although the two camps viewed each other with skepticism, both could agree that cleaning up the bases would be easier if backed by the president. With the White House expecting results, I knew I had some leverage.

Still, EPA officials were adamant that they didn't have the budget to adequately review DOD's cleanup efforts in the time required by the "fast-track" plan. So I offered to give EPA $7 million per year from DOD's budget to speed up the process. It worked. By the time I left

Figure 2-2. Secretary of Defense William Perry at a recycling ceremony at the Pentagon, December 1994 (photo credit: author's personal collection)

DOD in 2001, 80 percent of the property on the closed bases was ready to be transferred and reused. This partnership also helped build trust with EPA so that we could address other thorny environmental issues. Carol Browner, the EPA administrator, remarked at a public event, "Sherri Goodman has a lot of environmental problems at DOD, but the good thing about the military is that they know how to get things done."

## The Buck Stops Here

Brokering the compromise between EPA and DOD was an important step in accelerating cleanups of closing bases, but it was just the first one. Throughout my years at DOD, I faced relentless pressure to ensure that work was performed faster, cheaper, and better. Congress wanted cleanup to happen faster so that members could address their constituents' complaints; most of DOD wanted cleanup to be cheaper so that money could be spent on other, higher-priority matters; and local citizens wanted a more inclusive process that accounted for their concerns.

I often felt like a spinning top—trying to assure Congress that we were indeed moving faster and that yes, we deserved the funding we had requested; trying to stave off the green-eyed budgeteers at DOD who wanted to move the cleanup funds to other accounts; and listening to citizens, factoring their concerns into the design of government programs.

Nowhere did these tensions come together in a greater crescendo than at the Massachusetts Military Reservation (MMR) on Cape Cod. Surrounded by four shoreline towns, MMR had multiple military "tenants," from the US Coast Guard to the US Air Force to the Massachusetts National Guard and Reserve, each of which controlled different sections of the base. Each conducted its own mission, managed its own staff, and made its own mess.

By the early 1990s, industrial activities at the base had created plumes of contaminants that were migrating into the sandy soil of Cape Cod and threatening the local water supply. Multiple military services were responsible, but since no one service "owned" the base, none had stepped up to lead. Instead, there was a lot of finger-pointing. Local citizens were angry, and regulators at both EPA and the Massachusetts Department of Environmental Protection were exasperated.

Back at the Pentagon, I inquired which military service had the lead for cleaning up the base. No service raised its hand. The air force knew it had to deal with the pollution it had created, but it didn't always have the same approach as the army and the Massachusetts Guard. Local citizens and selectmen of the four surrounding towns felt they couldn't get straight answers from DOD. And when they did get an answer, they often didn't trust it. So they went to their elected representatives, including Senators Ted Kennedy and John Kerry and Congressman Gerry Studds, who was succeeded by Bill Delahunt in 1997.

I realized quickly that I would have to resolve this issue or end up taking the blame. I was emboldened by my roots on Cape Cod—I grew up walking its beaches and sailing its waters, and my parents had retired there. And I knew from my experience on the Senate Armed Services Committee that a powerful politician can move a recalcitrant bureaucracy if they know which levers to push. Senator Kennedy, chairman of the Armed Services Subcommittee on Seapower, knew how to get the Pentagon's attention. I took a calculated risk that if I committed DOD to spearheading the cleanup, I could get the military departments, with Congress's support, to follow through.

In the summer of 1994, I announced that DOD would coordinate and fund a plan to clean up the plumes that were threatening the Cape's only aquifer. At a local press conference on July 8, 1994, the Massachusetts congressional delegation, state and federal environmental officials, and local citizens each took credit for getting DOD to act. I did get some recognition from Congressman Studds, who said I had a "marvelous understanding of the threat posed by the plumes."[8]

The irony was that on the very same day the MMR cleanup plan was announced, I was defending to conservative critics why DOD was engaged at all in environmental matters. The *Wall Street Journal* ran an editorial on May 24, 1994, titled "The Army's Latest Foe," arguing that engaging in environmental cleanup was at odds with maintaining a military ready to fight. I responded in a letter coauthored with the vice chairman of the Joint Chiefs of Staff, Admiral Bill Owens, in which we wrote: "First, we are funding environmental cleanup and military operations and training separately. . . . Second, we've learned the hard way that, as a practical matter, the readiness of forces can be endangered unless the department complies with environmental laws."[9]

Responding to this type of criticism by the *Wall Street Journal* and similar organizations was a needle I had to thread throughout my time at the Pentagon. There always seemed to be a faction who saw environmental stewardship and military readiness as opposing forces, instead of two sides of the same coin. Thanks to leaders such as Admiral Owens, however, we were beginning to make progress.

By 1997, I had established a Joint Program Office for MMR, requiring the multiple military "tenants" to work together and communicate with one voice to the community. I also committed to making three million gallons of clean drinking water available daily to the Upper Cape within three years. The project involved three new wells and eleven miles of pipe carrying the water to each of the four towns on the Upper Cape—Bourne, Mashpee, Falmouth, and Sandwich. And to keep lead from bullets on the training range from leaching into the water supply, DOD worked to develop "green bullets," made of tungsten.

When I made my last official visit to MMR in 2000, I was able to observe the construction of the large pumps that would clean the water.[10] Local water officials heralded me as their heroine and gave me the perfect Cape Cod sendoff: a clambake, complete with lobster, steamers, and corn.

## Listening to Citizens

On Cape Cod and elsewhere, I learned quickly that some of my power within the Pentagon actually came from the outside: from citizens and their representatives demanding change in DOD's environmental practices. From California to the Carolinas, communities were complaining that DOD needed to clean up its mess.[11] They wanted to know more about what the military was doing on its bases, and they wanted to be part of the process.

In 1991, before I arrived at the Pentagon, Congress had created the Defense Environmental Response Task Force to examine environmental issues associated with military base closures and cleanup.[12] The task force included not only DOD and EPA officials but also representatives from the US Department of Justice, the National Governors Association, state attorneys general offices (which mostly sued the military), and various public interest groups. By assembling this diverse crew, Congress was making clear to the Pentagon that it wanted other voices to be heard on the issue of military cleanup. The environmental regulators and citizen stakeholders would press for more openness in military decision-making and more attention to local concerns.

The task force became my vehicle both for building partnerships with citizens who were critical of DOD's cleanup to date and for developing support within the military for a more rigorous and responsive cleanup program. We convened public meetings with citizens and stakeholders more than a dozen times during five years.[13] I traveled around the country with the task force, hearing testimony from concerned community members. We heard a lot of dissatisfaction. In San Antonio, Texas, residents around Kelly Air Force Base, which had closed in 1995, rightfully gave us an earful about how their water had been contaminated by toxic chemicals that had leached off base.[14]

After the first few meetings, I realized DOD needed to engage concerned citizens more regularly at the local level. The Defense Environmental Response Task Force was valuable, but it wasn't enough to handle continuous local feedback about decisions made at all the bases. So I called for the creation of DOD Restoration Advisory Boards (RABs).[15] Each military base with a cleanup program could form an RAB that included both DOD and local stakeholders. Importantly, the RABs would be cochaired by both the installation commander and a community representative. Native American tribal nations, whose land has frequently been used for military training ranges, were able to establish their own RABs.

One key citizen representative to DOD was Lenny Siegel, a mild-mannered environmental activist from Mountain View, California. Siegel organized his community to push for faster cleanup of Moffett Field Naval Air Station in Silicon Valley, which had closed in 1994. When I invited him to meet with me in the Pentagon in the early 1990s, he told me it was his first time in the five-sided building on the Potomac. After all, defense officials did not typically invite in citizen activists, who they tended to view as either radical or an enemy. Lenny and I developed a good working relationship, and I knew I could rely on him to be fair and factual. He would go on to win an award from EPA in 2011 recognizing his efforts to advance cleanup and preserve community values during the closure of military bases.[16]

Not all military bases could be cleaned up completely. Some would never be safe for a childcare center, school, or similar facility. Old training ranges with unexploded ordnance, for example, could not be developed.[17] And yet these lands are still valuable and often are well suited to become wildlife refuges. From Fort Ord in Monterey, California, to Rocky Mountain Arsenal National Wildlife Refuge outside Denver, Colorado, many former military bases have become "islands of nature" protecting endangered species and preserving wilderness.

## Tanks and Tortoises

The military's environmental awakening wasn't only about cleaning up pollution. It was also about preserving natural areas in and around military bases and training ranges. During the 1990s, I often found myself at the center of disputes about training and endangered species. Could our service members train in a way that wouldn't threaten ecosystems?

Most Americans don't know that DOD is the nation's second-largest landowner, managing over thirty million acres across all fifty states and territories. This land includes valuable habitats, from deserts to old-growth forests. As I told the *New York Times* in early 1996, military bases offer "the last, best hope for protecting certain kinds of ecosystems."[18] But to preserve them while meeting the military's mission to train troops, I would need allies at the highest levels of the military. I found one in General Gordon R. Sullivan.

General "Sully" Sullivan served as the thirty-second chief of staff of the US Army and was a father figure to the soldiers he led. With a wide smile and a twinkle in his eye, he evoked the regular-guy, G.I. Joe image of his generation. Sully was born in Boston and raised in Quincy, Massachusetts, a city that has produced more four-star officers than any other in the United States.

He dated his formative experience on the environment to his service at Fort Riley in Kansas, where he was in charge of range operations and training before becoming commanding general. He told me that during that time, he was "driven" to protect the bald eagles that nested on the range and that "the soldiers respected the eagle—they're not going to touch it or disrupt it."[19] Indeed, as of 2022, Fort Riley was within the historical range of thirteen federal and state threatened and endangered species, including bald eagles and whooping cranes.[20]

By the time I met General Sullivan in the early 1990s, he was in the midst of resetting the US Army after the Cold War. Sully led the transition of the service from large, mechanized battalions prepared to fight the Soviets in a major European land war to a lighter, more agile force shaped by the conflicts in the Persian Gulf and Panama in the late 1980s and early 1990s. During this period, he also helped the army navigate significant environmental challenges. Suffice to say, it was not an easy process, with several black eyes along the way. At Aberdeen Proving Ground in Maryland, where the army was developing and testing chemical weapons, service members dumped lethal and carcinogenic chemicals down drains and into a creek, leading to a fish kill. In 1989, federal regulators shuttered this laboratory, and in 1994 they fined the army for damages. An even greater wake-up call came in 1991 when the US Fish and Wildlife Service effectively shut down parts of the training range at Fort Bragg in North Carolina (now Fort Liberty) for failure to protect the nesting grounds of the endangered red-cockaded woodpecker.

The army's initial response was hostile: "The base is supposed to be a battlefield, and you don't have a battlefield anymore," grumbled one military specialist. "It's against all kinds of military doctrine."[21] But Sully acknowledged the problem, and the army began to take steps to accommodate the RCW, as the endangered bird came to be known. At first, they simply cordoned off the trees where the birds nested, but that meant the army could no longer train as before.

Nearby, Marine Corps Base Camp Lejeune was dealing with the same concern. The marines' training ranges would also be shuttered if they didn't protect the longleaf pine where the RCW lived. So they devised a different strategy: the trees became "realistic obstacles" around which marines would have to maneuver without firing. In fact, the marines even ran an environmental campaign about the RCW, creating a poster with a marine in camouflage uniform holding a rifle next to a longleaf pine with a bird perched in its nest. (It is one of my most cherished DOD keepsakes, signed by then commandant Charles "Chuck" Krulak.)

The army realized it could change its training practices to protect the home of the endangered bird. Former Fort Bragg official Mike Lynch recalled that initially "the woodpeckers were a problem.... We couldn't sit in a room for more than 20 minutes without getting into a heated debate." The biologists and the soldiers were speaking different languages. But eventually, Lynch, like General Sullivan, came to see they had common interests. "We wanted good forests. We wanted places ... that were going to be here in 100 years."[22]

Both the US Army and the US Marine Corps altered their routines to make the longleaf pine a realistic obstacle *in* training rather than an obstacle *to* training. Not only did DOD change its practices, it also began actively working with conservation groups, land trusts, state and local governments, and private landowners to conserve habitat both on and off military bases.[23] These partnerships and conservation easements have become widely used tools both to protect bases from encroaching development and to preserve regions' natural habitats. In essence, both the military and nature benefited from this arrangement.

Across the country from Fort Bragg, deep in the Mojave Desert, the army and the marine corps conduct some of their most important training before going to war. On the wide-open desert at Fort Irwin National Training Center, General George Patton's tank tracks are still visible. America's troops trained there before deploying to the Gulf War. At the marine corps' training center at Twentynine Palms, California, troops trained for service in both Iraq and Afghanistan. Slow to grow and easy to disrupt, the Mojave Desert is also home to the endangered desert tortoise.

In 2012, the army's website proclaimed that "Fort Irwin is home not only to great Soldiers, Family Members, Department of the Army Civilians, and Contractors, it is also home to Gopherus Agassizii: the Desert Tortoise."[24] Back in the 1990s, however, these efforts were just getting started. Indeed, my first congressional testimony at DOD was about the military's request for land in California to be withdrawn from public use for military training. The politically important state was willing—but only if the Mojave Desert was protected. In my testimony, I stated, "As we change our paradigm to meet new threats, I am here today to support legislation that protects the California Desert and to tell you that the President, Vice President, and Secretary Aspin believe that environmental security is vital to national security."[25]

While my words sounded good in theory, they were, of course, harder to put into practice. Some of the local advocacy groups didn't trust the military. The largest landowner in the Mojave Desert, the Bureau of Land Management, had its own challenges in mediating conflicts between hunters and drivers of all-terrain vehicles, who wanted more recreational access, and environmentalists, who wanted as few footprints as possible on a fragile ecosystem. Meanwhile, the army's new technology for tanks and armored vehicles meant it needed ever-larger areas in which to train. The army proposed expanding its training range at Fort Irwin, which could affect a local road and possible desert tortoise habitat.

What's an army chief to do? General Sullivan knew instinctively that he needed to showcase the army's dual commitment to training and to the environment. For Earth Day in May 1994, Sully and I traveled to Fort Irwin to elevate the work of the base's natural resource managers. They showed us how the tortoise habitat was protected and how the nests with eggs were covered with chicken wire to protect them from prying ravens. As we posed for photos with a tortoise, Sully took it all in, as happy to be with the biologists as with the soldiers.

Several years later, Fort Irwin and other military bases in the Mojave Desert received an award for their work on cooperative management of the tortoise and protection of the ecosystem.

**Cooperation Abroad**

Cleaning up bases and conserving endangered species was just the beginning of DOD's evolving approach to the environment. In the 1990s, we were deeply engaged in working with militaries overseas, both to extend best environmental practices and to thaw relations with countries worldwide after the Cold War. During those years, I frequently gave speeches and wrote articles about "engaging military ministries across the globe in an effort to create conditions of peace, not just conditions of conflict." The effort was akin to the Marshall Plan's objectives of rebuilding Europe after World War II—a way to "build trust and understanding among the militaries of the region."[26]

In fact, DOD's cooperation with foreign militaries in the 1990s was a key component of America's broad national security objectives of this era, to extend peace and prosperity to the new democracies in central and eastern Europe and to develop constructive relationships with Russia and the other newly independent former Soviet states. In Poland,

Hungary, the Czech Republic, Ukraine, Belarus, Kazakhstan, Georgia, and Russia, we worked with our foreign counterparts in conducting environmental education and training, sharing remediation methods, and showcasing environmental technology.

Not only did we share techniques and tools; we also modeled the process of interagency collaboration by bringing other US agencies, such as EPA, the US Department of Energy, and the US Department of Commerce, into these efforts. I felt a genuine spirit of cooperation both within our own government and with our allies, a sense that the armed forces need not be environmental laggards but could become true leaders. Speaking at Wright-Patterson Air Force Base, I observed: "Our military has become very conscious of the need to be good stewards. . . . Practices 10 years ago, even 5 years ago, are not the same as they are today. The attitude is not to leave this legacy for another generation."[27]

CHAPTER 3

# Generals and Admirals Battle Climate Change

WHEN IT CAME TO ENVIRONMENTAL PROTECTION—both inside and beyond the military—the 1990s were a time of great promise but also great uncertainty and conflict. While our armed forces were cooperating with their foreign counterparts on environmental issues, the United States was debating our position on the Kyoto Protocol, the first international treaty to set legally binding targets for individual countries to cut greenhouse gas emissions.[1] Before agreeing to the targets, nations would have to wrestle internally with how much they were willing to cut and how to go about it.

Leading up to the landmark conference in Kyoto, Japan, in 1997, debate was fierce between Americans who saw the Kyoto Protocol as a vital step in combating the dangers of climate change and those who worried that slashing carbon dioxide would also mean harming our economy—and our security. The US Department of Defense (DOD), as the federal government's biggest energy user (72 percent at the time)[2] and therefore greatest contributor to emissions, would have a small but important role in that debate. And I, as DOD's senior representative on climate matters, would make the department's case to the president and his key advisors.

The domestic battles surrounding the Kyoto Protocol were shaped by both the scientific understanding of climate change at that time and economic pressures after the Cold War. First, while most scientists agreed that rising greenhouse gas emissions would dramatically affect global temperatures, the scientific discourse was about "potential" or "projected" climate change. Many observers believed that global warming would happen in the future, if at all. Americans had not yet confronted the daily visible evidence of a warming planet: persistent wildfires, worsening storms, accelerating sea level rise and coastal erosion, and persistent drought along with torrential rains. Terms such as "bomb cyclone" and "derecho" had not entered the common lexicon.[3] Moreover, no US law required reduction of greenhouse gases.

Second, the politics of the environment were becoming increasingly partisan. While the 1996 elections had kept President Bill Clinton and Vice President Al Gore in office, Republican congressman Newt Gingrich was speaker of the US House of Representatives, and the GOP had won control of the US Senate. The Republican Congress opposed much of the Clinton administration's environmental agenda, particularly on climate change. But even Democrats were loath to commit the United States to reductions in greenhouse gas emissions as long as China and other developing countries were not fully on board. In July 1997, the Senate introduced the Byrd-Hagel Resolution, stating that the United States should not sign a climate treaty without mandates for developing nations (mainly China and India) because doing so "would result in serious harm to the economy of the United States."[4] The resolution was cosponsored by Senator Robert Byrd of West Virginia (the most senior Democratic senator at the time) and Senator Chuck Hagel of Nebraska, a Republican. It passed 95–0.

Third, military leaders worried that any agreement at Kyoto would put even greater strain on a defense budget that had already been cut dramatically. The United States had reduced its defense budget by 38 percent since 1985, at the end of the Cold War, and had shrunk its force posture (military capability) by 33 percent and its modernization programs by a whopping 63 percent. While the end of the Cold War had produced this "peace dividend"—meaning the economic benefit of a decrease in defense spending—there was concern within the armed services committees that the United States was not investing enough in its own security and therefore undermining the readiness of its forces to respond to future threats. I noted in a speech to a DOD environmental conference at the time that Congress was more under the sway of the "budget hawks" than of the "defense hawks" and thus not likely to see an immediate increase in its modernization budget.[5] With defense budgets already squeezed, therefore, any additional priority, such as a response to the threat of climate change, would not be viewed favorably by the armed services committees.

These scientific, political, and defense challenges formed the backdrop to a remarkable statement President Clinton's advisors made in a memo about decisions he needed to make in advance of the Kyoto negotiations: "*The decisions identified in this memo may be some of the most important and difficult of your Presidency* [emphasis added]. There does not appear to be any way fully to reconcile four constraints we have previously identified to you—the environmental imperative, diplomatic realities, economic costs, and political opposition from business, labor and the Hill."[6] Yet reconciling all of these pressures was the goal as the Clinton administration prepared to head to Japan.

DOD was a small dog in this fight. In the advisors' lengthy memo to the president—fourteen pages, with many decision boxes to check—national security issues appeared on the very last page. Still, the military had a role to play, and my job was to coordinate DOD's position as part of the overall US strategy. The central question was, How could the military reduce emissions without sacrificing security?

At the time, cutting greenhouse gas emissions translated directly into cutting energy use. And for the military, cutting energy use had a direct impact on the military mission. (We will see later that the Gordian knot that tied fuel to military effectiveness has now been severed.) When the military services were asked, during the preparation for the negotiations, how cutting national emissions would affect their operations, they came back with hair-raising numbers: a 10 percent reduction in DOD emissions would translate into a "cut of 328,000 miles a year for Army tank training, a loss of 2,000 steaming days a year for the Navy and a reduction of 210,000 flying hours for the Air Force."[7] It's fair to say that making those changes would affect every dimension of the military. If the United States agreed to an overall reduction goal, emissions would need to be reduced from every military base; every vehicle; every tank, ship, airplane; every training activity; and potentially every military operation.

This was not a matter of modifying training drills to accommodate tortoises or cleaning up military bases. In those cases, the military was mandated by federal and state environmental laws to protect human health and natural resources—and there was funding to do so, not to mention serious legal consequences for failing to comply. Even if environmental concerns were not their first priority, leaders at DOD could clearly see the benefits of healthier air and water for military service members and their families. And thanks to General Gordon "Sully" Sullivan and others, there was a growing appreciation for stewardship of natural resources. Perhaps most importantly, cleaning up bases and protecting local environments did not require the military forces to substantially change their operations.

The scope of climate change made these previous problems look easy. If cutting greenhouse gas emissions included emissions from all military activities, would that mean that fighting a war would be constrained by greenhouse gas limitations? The United States and our allies in the North Atlantic Treaty Organization could not accept limits on the use of force that would not apply to our adversaries, including China and Russia. With fighters, bombers, nuclear submarines, aircraft carriers, tanks, and troops operating all over the world, tracking emissions associated with military operations was no easy task.

As the person charged with crafting DOD's strategy on climate change, I struggled with how to demonstrate that the department was serious about climate change while neither compromising military readiness nor constraining the use of forces, almost all of which operated on fossil fuels at the time.

My team and I worked hard to meet the challenge: developing a rigorous climate change program that would not jeopardize the readiness of our fighting forces or our ability to conduct military operations around the world. We concluded that we could more readily reduce emissions from US bases and from nontactical vehicles (cars and trucks) than we could from forces deployed around the world. As a result, we requested an exemption for military operations, reasoning that flying planes or steaming ships should not be constrained by limits on the amount of fuel they could use.[8] The uniformed military felt particularly strongly that this was a question of basic readiness. Chuck Wald, who at the time was brigadier general of the US Air Force and would later become an outspoken proponent of climate action, stated, "A reduction of 20 percent of fossil fuel emissions would be a killer."[9]

Would I have preferred not to need a national security exemption? Absolutely. I recognized that this position was disappointing to environmental advocates, and I shared their belief in the importance of addressing climate change. Yet I also knew we needed more time to find ways to lower the energy needed for military operations without threatening their effectiveness. Over the long term, the solutions would come down to reducing the logistics burden associated with fuel consumption. In 1997, we understood that was the objective, but we didn't know yet how to get there. The military's concerns about emissions limits prevailed with the White House and Secretary of State Madeleine Albright, who also supported an exemption.[10] Secretary Albright was a proponent of a muscular foreign policy and did not want to constrain the military's ability to back up diplomatic agreements with force.

The president's final decision, as reflected in a National Security Council memorandum, stated, "Measures intended to promote reductions in emissions of greenhouse gasses shall not impair or adversely affect military operations and training."[11] Even so, DOD could claim that it had met the existing goal of reducing emissions by 7 percent below 1990 levels by 2008–12. In fact, DOD had already reduced its emissions by 25 percent from 1990 levels, attributable in large measure to base closures during that era but also to improved energy efficiency. Indeed, DOD had already reduced its facility energy consumption by 17 percent per square foot since 1990.

We devoted much of 1998 to developing DOD's first ever climate change strategy, focused primarily on emissions reductions. It had four components: First, continue reducing energy use in facilities and nontactical vehicles. Second, develop weapons systems technologies that improve performance while reducing energy use. Third, create the first inventory of DOD greenhouse gas emissions. Fourth, "educate and cooperate."[12] The concluding slide of my briefing, "Department of Defense Climate Change Programs," stated enthusiastically "IT CAN BE DONE!" and noted that "DOD will continue to be a leader in combating climate change" and "government and industry have a stewardship obligation to the taxpayer."[13]

In the end, however, with the Byrd-Hagel Resolution serving as a sword of Damocles hanging over any climate treaty, the US Senate did not ratify the Kyoto Protocol, and the exemption DOD sought was never invoked. Yet even without limits mandated by the treaty, the debate prompted DOD to rethink the connection between energy use and military performance. To compel a rethinking of energy use for military operations, one of my last, and maybe more important, efforts as deputy undersecretary was asking DOD's Defense Science Board, along with DOD's director of research and engineering, to conduct a study titled "More Capable Warfighting through Reduced Fuel Burden."[14]

The Defense Science Board, composed of distinguished academics, technologists, and former military officers, is known for taking on hard science and technology problems. This study, led by an experienced group of energy experts and retired military leaders, including the director of the US Department of Energy (DOE)'s National Renewable Energy Laboratory, Vice Admiral Dick Truly, detailed technologies and operational techniques DOD could adopt to cut energy use in weapons systems. Admiral Truly told me, reflecting on this study, that his task force had no trouble finding examples of "gas-guzzling weapons systems," but the problem was that energy efficiency was "absent in the original requirements. . . . We set out to change that," he said.[15]

EPA Administrator Carol Browner recognized this effort, and DOD's climate leadership more broadly, in an award I received in October 2000:

> Sherri Goodman, the most senior US DoD environmental policy-maker, has helped DOD reduce its overall greenhouse gas emissions by approximately 26 percent while improving military capability. Under her leadership, DOD has initiated a program to understand how climate change will affect future military operations and sponsored a study by the Defense Science Board of how more fuel-efficient weapons systems will improve warfighting capability and reduce operating costs.

She also initiated a program to determine baseline emissions and instituted measures to continuously monitor carbon emissions and sinks.[16]

As it turned out, the Defense Science Board's study was released just prior to the tragic terror events of September 11, 2001. The US military would soon find our sons and daughters in a war on the other side of the globe, with many lives lost while guarding or transporting fuel. The wars in Iraq and Afghanistan would become the catalyst for DOD to once again pay more attention to energy and, ultimately, climate change.

## A New Kind of Mission

As the wars dragged on in Iraq and Afghanistan through the early aughts, with almost three hundred thousand US troops deployed, the US military began to see the human costs of climate extremes and protecting fuel and water supply lines. In drought-parched Afghanistan, one soldier was killed for every twenty-four convoys to resupply fuel or water, while troops baked in 115-degree temperatures.[17] Water scarcity was projected to cripple 40 percent of the world's nations in coming decades, leading to more conflict.[18]

And yet the national conversation about climate change at that time was primarily about polar bears or a polarized debate about Al Gore's *Inconvenient Truth*, which was either gospel or heresy, depending on your point of view.[19] Renowned climate scientist James Hansen had testified about the risks of climate change as early as 1988, but his alarm had not broken through to a broad section of the American public.[20] Climate change was associated with so-called tree huggers, despite President George H. W. Bush's earlier attempts at the Earth Summit in Rio de Janeiro to frame the issue as a global challenge and efforts by some analysts in the armed services to warn about its serious security consequences (such as a 1990 report by the US Naval War College titled "Global Climate Change Implications for the United States Navy").[21] The oil and gas industry had been sowing seeds of scientific uncertainty and economic harm for over a decade.[22] Many moderates and conservatives didn't identify with climate action, nor did their elected officials. Major climate bills were stuck in Congress. From the high point of bipartisan action in the George H. W. Bush administration, environmental issues had become a political battleground by the latter half of the 1990s, and they stayed that way into the first decade of the new century.

My own work on climate had also taken a pause. On January 19, 2001, I left the Pentagon after almost eight years of service as deputy undersecretary of defense (environmental security). A new president would take office the following day. At my farewell, my office mates and boss feted me, eight and a half months pregnant, with gag gifts including a plaque inscribed Mother of Environmental Security.

Keeping to my adage that for parents of young children, work is a refuge from the chaos at home, I joined the Center for Naval Analyses (CNA) when I left the Pentagon, thanks once again to Bob Murray, CNA's president. Founded in 1942 by Massachusetts Institute of Technology scientists and engineers to help the US Navy track German U-boat submarines during World War II, CNA is often thought of as the birthplace of military operations research. From this one assignment, CNA ultimately became the research and analysis arm, along with its sibling institutions the RAND Corporation and the Institute for Defense Analyses, of the US military and intelligence community. After joining CNA in 2001 as a part-time senior fellow, I acquired more duties over time, including general counsel, corporate secretary, and senior vice president. With multiple hats at CNA and three young kids, I had my hands full. I didn't realize that one of my most important challenges on climate change was yet to come.

In 2006, I spoke with a group of foundations about a question that was in many ways the elephant in the room during the Kyoto climate negotiations: What are the national security implications of climate change? I recognized the dearth of research on this issue, and the foundations believed that examining it could help debunk the common belief that only environmentalists care about climate change. As Lee Wasserman (no relation) of the Rockefeller Family Fund, the original funder of our study, told me, they were trying "to figure out who could be [their] Sherpa to the military world," and CNA and I were at the top of their list.[23] Given my experience in DOD, I knew that environmental forces were affecting the military and that the military could positively shape the direction of US environmental policy. I also knew the power of the respected voice of military leaders.

I invited a dozen senior military leaders I had worked with in the 1990s, now retired, to join me on this climate journey. Even though I knew most of them well from my eight years at the Pentagon, I had to screw up my courage to ask them to risk their hard-earned reputations on a subject that was not in their traditional military wheelhouse. Most of them were initially skeptical. "I'm not a scientist. I'm a warfighter. I don't know anything about climate change." Yes, I acknowledged. Yet I knew that to become four-star generals and admirals, they had been willing to address the complex risks facing America and our military forces. Plus, we had a personal bond that stretched back to our time together in the Pentagon and to our work together when they were at the pinnacle of their military careers. A common military refrain, made popular by the Stephen Covey book of that era, is that "decisions move at the speed of trust."[24] I was now asking them to trust me on a very different kind of mission.

We became the CNA Military Advisory Board, with eleven retired three- and four-star military officers, all deeply respected for their military service and leadership. Among the founding members were former chief of staff of the US Army General Gordon Sullivan, former commander of the US Central Command General Anthony Zinni, former commander of the Pacific Command Admiral Joseph Prueher, and former director of the Naval Nuclear Propulsion Program Admiral Frank L. "Skip" Bowman. General Paul Kern, former chief of army logistics and one of the original board members, told me, "You built a lot of confidence in the military folks that this effort was not going to undermine the ability for the military to do their job but in fact, you would protect it."

Most of these distinguished warfighters had, in their own way, dealt with environmental concerns during military planning, whether to engage allies in the Pacific region, the Middle East, or the former Soviet Union, or to reduce the boot print of military training. Some, however, had tried to stay as far away from tree-hugger issues as possible. Vice Admiral Truly, who served as test pilot in the 1960s and astronaut aboard the space shuttle *Columbia* in 1981, told me, "I was a child of the DOD culture, which in my view, pretty much ignored environmental considerations."[25] Truly's views would change so much that he would later lead DOE's National Renewable Energy Laboratory and would co-lead the previously mentioned study on fuel-efficient weapons systems.

Truly was not alone in his evolution from climate skeptic to climate hawk. General Sullivan told Congress, "When I was asked to be on the Military Advisory Board, I was both pleased and skeptical." After fifty years of military service, he expressed concern that the sacrifices he and his soldiers made for America were "being overtaken by a much more powerful and significant challenge to the health and well-being of my 3 grandchildren and their children."[26] General Sullivan was an accurate barometer of the military leaders as a whole, who ultimately overcame their reservations and became convinced by the evidence that climate change would have direct, significant impacts on the nation's national security in the coming years.

And yet this change in thinking was neither an easy nor a straightforward process. During our year together, each of the leaders struggled with how to think about climate change in the context of their own military experience. These struggles took three forms: First, how do we reconcile scientific uncertainty about climate projections with the warfighter's need to assess risk with incomplete information? Second, how does climate change intersect with geopolitical risks of terrorism, violent extremist organizations, failed states, and great powers, such as Russia and China? Third, how did each military leader's personal experiences help him understand the threat posed by climate change?

**The Fog of War**

When the CNA Military Advisory Board began its mission, members needed to learn enough about climate science to apply this knowledge to their military experience as warfighters, planners, and leaders. Who better to provide those lessons than the world's leading climate scientists? Our first briefer was renowned climate scientist James Hansen, lead climate scientist and director of the National Aeronautics and Space Administration's Goddard Institute for Space Studies. Hansen had been testifying to Congress on climate science since the 1980s. Our second briefing was from Anthony Janetos, another leading climate scientist, then at the H. John Heinz III Center for Science, Economics and the Environment.

Both Hansen and Janetos provided extensive evidence on the science and trends in projected climate change. They were peppered with questions by the group of generals and admirals, who were accustomed to throwing hardballs at their briefers. How else had they gotten to their esteemed positions? Vice Admiral Truly recalled that at the end of this first session, he was struck by the "divergence of views [among our members] regarding whether climate change was real, and the group decided that they would take no stand on this question. They would leave it to the climate scientists."[27]

When I invited James Hansen to make this presentation, I specifically asked him to address scientific uncertainty regarding the effects of global warming. His briefing illustrated that regardless of the sophistication of the climate models, predictions are not perfect. We cannot know for sure exactly how fast the planet will warm and what the exact consequences will be. For some people, this uncertainty was a reason to delay action on climate change. If we can't know how quickly temperatures will rise, why do anything now? What if some warming is caused by a natural evolution of Earth systems and is not human-induced warming from coal, oil, gas, and other fossil fuels?

But this was not how the military leaders thought about uncertainty. As warfighters, they were accustomed to having to make important decisions, often life-and-death decisions, without complete information. Indeed, the very term "fog of war" is defined as "the uncertainty in situational awareness experienced by participants in military operations."[28]

Consider how General Dwight Eisenhower prepared for the famous D-Day landings in 1944. He had assembled the greatest array of military forces in history to invade Normandy, France, and stop the Nazis. Yet when exactly those forces should strike would depend on the weather. The invasion was tentatively scheduled for June 5, 1944, but on the morning beforehand, meteorologists disagreed about the forecast. They knew they would have to recommend a landing date and time to General Eisenhower in the face of significant uncertainty.

These meteorologists were operating without the modern technologies—satellites, weather radar, computer models, and instant communications—that today's forecasters take for granted. Instead, they relied mainly on surface observations from uniformed military personnel and civilians in Britain and Western Europe and a few military observers at sea. Predicting the exact timing, track, and strength of developing storms in the Atlantic Ocean was nearly impossible.

On the afternoon of June 4, when the weather began to deteriorate as the first storm approached, British Royal Air Force chief meteorologist and group captain James Stagg noticed an observation from a single ship stationed six hundred miles west of Ireland that reported a rise in barometric pressure. Stagg deduced that there could be a break in the weather on June 6.

That forecast turned out to be a pivotal moment in world history. Had Stagg been wrong, the lives of thousands of men and massive amounts of equipment could have been lost. General Eisenhower relied on the prediction, made with incomplete information, and moved the landing date to June 6, the day that turned the tide of World War II for the Allies.

Dealing with uncertainty is necessary not only during specific battles but also when planning long-term military strategy. In fact, the entire Cold War might be considered an exercise in preparing for the unknown. During those years, much of America's defense efforts focused on preventing a Soviet nuclear missile attack, what we referred to as a "bolt out of the blue." Our effort to avoid such an unlikely event was the central organizing principle for our Cold War defense strategies.

Since the late 1940s, the United States has devoted a significant portion of its gross domestic product to buying nuclear-armed missiles, aircraft, and submarines, all in the hope of never needing to use them. The strategy of nuclear deterrence is like an insurance policy: building up America's overwhelming nuclear capability so that the Soviet Union, and today Russia, China, North Korea, and Iran, won't be tempted to use their own.

A nuclear strike is a low-probability, high-consequence event—one with such devastating human consequences that it's worth spending much of America's treasure to prevent it. If a nuclear attack by the Soviet Union was a low-probability event with extremely high consequences, then climate change, General Sullivan said, was exactly the opposite: "We have a catastrophic event that appears to be inevitable. And the challenge is to stabilize things—to stabilize carbon in the atmosphere. Back then, the challenge was to *stop* a particular action. Now, the challenge is to *inspire* a particular action [emphasis added]."[29] That action is reducing the emissions that lead to runaway climate change.

Admiral Skip Bowman put it another way, from the viewpoint of a nuclear submariner: "We operate in an unforgiving environment—the seawater is constantly trying to get at us. And nuclear reactors are an unforgiving technology if you don't do all the proper things."[30] As a result, the navy institutes rigorous procedures to guard against failure of the nuclear technology on which the lives of these submariners depend. "We should begin planning for a similar approach in dealing with potential climate change effects on our national security," Admiral Bowman concluded.

Beyond uncertainty, the military leaders grappled with how climate change affected the stability of various regions of the world. In fact, destabilization—often caused by terrorism, violent extremist organizations, governance failure, and competition between major powers—goes to the heart of national security risks. One reason human civilizations have grown and flourished over the past five millennia is that the world's climate has been relatively stable. However, when environmental conditions deteriorate to the point that necessary resources are not available, societies can become stressed—even to the point of collapse.[31]

To evaluate how climate stressors were playing out around the planet, each general and admiral took responsibility for a specific region. (The following chapters examine these regional threats in detail.) The military leaders brought to the task the weight of their own experience in command, carefully assessing how climate had changed the war planning and military operations they led. For example, General Zinni, who led military operations in Iraq and Somalia, could see already that we needed to "look at how climate change effects could drive populations to migrate. Where do these people move? And what kinds of conflicts might result from their migration?"[32] General Zinni also observed that climate change is a "Petri dish for terrorism."[33]

The leaders' experiences with climate change were professional but also deeply personal. Each general and admiral had his own lens through which he saw climate threats. And that human connection was different for different people.

For General Sullivan, it was the awareness that his beloved maple trees of Vermont and New England might no longer be the defining feature of a future landscape, almost as if his best childhood memories were being taken from him. For General Zinni, who spent so much of his career conducting humanitarian relief missions in the Middle East and North Africa, it was a profound sense that the people in these regions would be even more vulnerable as water and food become scarcer with a worsening climate. For Vice Admiral Truly, it was his recollection of looking back at Earth from aboard the space shuttle *Columbia* and observing that "the only evidence of an atmosphere was a very thin line of color on the horizon. The image of that scene is burned into my memory," he said. In a way, each leader had his own aha moment that enabled him to rely not only on the evidence that had been presented but also on personal conviction, based on deep human experience.

## Threat Multiplier

After almost a year of study and deliberation, we convened as a group to hash out our findings. We were getting closer to consensus, but we weren't there yet. I was among the first to arrive at the Virginia think tank on that winter day in 2007. The space was cavernous, with the same drab gray rug and off-white walls as in many conference rooms in the Pentagon. A long, highly polished table anchored the room with heavy fake-leather conference chairs and a large screen at the front. Name cards had been carefully set out, reflecting the rank of each officer.

We had been meeting since 2006 at the CNA office to receive briefings from climate scientists and other experts. We had even traveled to London to get the most recent climate assessments from the world-renowned

British Meteorological Office and visited 10 Downing Street to meet with Prime Minister Tony Blair's top climate advisors. We had also met with industry leaders, including those at Walmart, who explained why reducing climate and energy risk was good for business.[34] Now, we had a deadline to meet. Soon, Congress would be considering the following year's budget, and we wanted our findings to be considered as part of the annual defense authorization bill. We also needed this information to be out in time for the United Nations Security Council's first session on climate security, in April 2007.

General Sullivan, former army chief of staff and by protocol the most senior of the many four-stars, chaired our decision meeting. Despite his vaunted status among this group, he had a folksy style. He would wait until everyone had spoken, calling on each of the raised placards around the room. "Okay, gang, we need to come together." He had been listening to the debate among the three- and four-stars in the room for months. Even after presentations by leading climate scientists, he and others had more questions. Some wanted to be sure we also heard from industry. So we did that too. After we'd been briefed, and briefed again, Sully sat straight up in his chair at the head of the long conference table.

"We seem to be standing by and, frankly, asking for perfectness in science," General Sullivan said. "Well, we know a great deal, and even with that, there is still uncertainty. But the trend line is clear. We never have one hundred percent certainty. We never have it. *If you wait until you have one hundred percent certainty, something bad is going to happen on the battlefield.*"[35]

His words landed like a tablet sent from above. They captured the essence of climate change for many of the generals and admirals in the room: Military decisions are always made in the fog of war. If we had perfect information, there would be no battle, only winners and losers. But because we never knew how well armed the opponent was, or exactly where the enemy had placed its battleships, warfighters and military planners always made decisions in the face of uncertainty. Uncertainty was not crippling. In fact, it was the very uncertainty of climate change that created the urgency for action.

As these stories accumulated, of warfighters who were making their own links from climate change to instability or from climate change to terrorism, I was thinking of how to explain this phenomenon to people who hadn't spent decades in the military. To make a difference, our ideas needed to resonate beyond the Pentagon yet use language that was familiar in defense circles and conveyed the appropriate gravitas. One common phrase in defense-speak is "force multiplier," which refers to a combination of factors that enables personnel or weapons to accomplish greater feats than without it. For example, a technology such as GPS is considered a force multiplier because it enables greater accuracy in military operations.

As I listened to the generals and admirals, it hit me. "How about the term 'threat multiplier'?" It stuck. The military leaders liked the term because it had a hard-defense edge. The communications team liked it because it was short and authentic. The opening letter of our report, signed by each member of the CNA Military Advisory Board, stated: "Climate change can act as a threat multiplier for instability in some of the most volatile regions of the world, and it presents significant national security challenges for the United States." "Threat multiplier" not only became our signature phrase; it was later used in numerous US, DOD, North Atlantic Treaty Organization, and UN reports, has been regularly cited by journalists, and has informed thinking and policy action on climate security over the many years since we gathered on that day in early 2007.[36] We did not imagine, when we raised our voices to proclaim this emerging threat, that it would accelerate with such fury in the coming decades and that the security consequences of climate change would become so critical. In the short term, the military leaders' message yielded some immediate, while incremental, progress.

On April 16, 2007, we held a press conference to call attention to the security threats presented by climate change. Just a day later, the UN Security Council held its first meeting on the issue.[37] British foreign secretary Margaret Beckett echoed our language, observing that climate change "is an issue which threatens the peace and security of the whole planet."[38] The United Kingdom's initiative on climate and security received a mixed reception at the United Nations.[39] European countries praised Britain, while developing nations accused it of trying to grab power. Meanwhile, the United States, deep in the wars in Iraq and Afghanistan, was keeping an arm's length from this international debate.

Still, there was some movement at home. That same week, the US Senate Committee on Armed Services, in an amendment offered by Senator John Warner of Virginia, a Republican, and Senator Hillary Clinton of New York, a Democrat, added our first recommendation to the annual defense authorization bill. The provision required DOD to consider the effects of climate change on military missions, including preparedness for natural disasters.

Meanwhile, over in the US House of Representatives, Congressman Edward Markey of Massachusetts, a Democrat, had just become chair of a new Select Committee on Energy Independence and Global Warming. He was eager to make his mark and elevate climate change to greater national attention, so he planned a series of hearings on the issue. He knew that if he began these hearings by focusing on the security implications of climate change, he could create some bipartisan support for what was becoming an increasingly polarized issue in the United States. His first witness was General Gordon Sullivan. Congressman Markey remarked, "Today's witnesses have been invited because they have spent their lives thinking about what must be done to defend our planet, and how the twin imperatives of defending our environment and defending our freedom have begun to merge into [a] single issue."[40]

A few weeks later, Admirals Prueher and Truly and General Chuck

Wald testified before the US Senate Committee on Foreign Relations, chaired by Senator Joe Biden of Delaware.[41] Over the next two years, at least five different congressional hearings would feature at least eight testimonies from various members of the CNA Military Advisory Board.

The term "threat multiplier" was also used in two milestone pieces of US legislation on the impacts of climate change on security: the Global Climate Change Security Oversight Act (2007) and the Lieberman-Warner Climate Security Act of 2008.[42] Both emphasized the role of changing weather patterns on global stability. Two years later, the American Clean Energy and Security Act of 2009 described climate change as a "potentially significant national and global security threat multiplier."[43]

Informed by the military leaders' recommendations, these landmark laws led to the creation of climate security policies within the US government, encouraging a systemic approach to multiple risks. In turn, DOD began to formally recognize climate risks, describing in its 2010 "Quadrennial Defense Review Report" the "significant geopolitical impacts" of climate change on security concerns, such as migrations and the weakening of fragile states. The document further argued that climate change "may act as an accelerant of instability or conflict."[44]

Even beyond the rarefied confines of certain Washington policy circles, the idea of climate security was gaining traction. In 2007, the *New York Times Magazine* called out our work in its Year in Ideas with the new idea of "climate conflicts," stating the following:

> It took years for a consensus on the existence and causes of climate change to emerge. But it took no time at all, it seems, for leaders around the world to latch onto the notion that global warming will bring war. In the spring, a report by retired U.S. generals and admirals called on Washington to incorporate climate change, especially its destabilizing effect on weak states, into the United States' national defense strategy.[45]

## The Creation of a Brand-New Field

Beyond its immediate effects, our 2007 report provided the seeds for a whole new field of study now commonly called climate security, which continued to develop in subsequent years. In 2011, two entrepreneurial policy leaders, Francesco Femia and Caitlin Werrell, launched the first stand-alone think tank dedicated to climate security, aptly called the Center for Climate and Security.[46] The organization would go on to establish the first community of practice on climate security in the United States, the Climate and Security Working Group, and, later, its public organ, the Climate and Security Advisory Group, which would produce annual policy recommendations for the US government endorsed by experts in security, military, intelligence, and climate security.[47] These efforts have proved crucial in shaping US policy on climate security over the past decade.[48] Important related efforts were led by Sharon Burke and Christine Parthemore at the Center for a New American Security's national security program, Kent Butts at the US Army War College, Rich Engel at the National Intelligence Council, and P. J. Simmons and Geoffrey Dabelko at the Woodrow Wilson International Center for Scholars.[49]

After 2007, the CNA Military Advisory Board produced seven more reports on the national security implications of climate change, energy, and water stress and scarcity. Of equal, if not greater, importance, the scores of generals and admirals who have served on the CNA Military Advisory Board have become national and international communicators on climate, energy, and national security. They have spoken at Rotary Clubs and local universities, from Memphis to Miami, from Dallas to Detroit, and beyond. They have continued to testify before Congress and make direct appeals to the president to take climate security seriously.[50] And they have spoken before such organizations as Conservatives for Clean Energy and the Evangelical Environmental Network.[51]

Senator John Warner, a Virginia Republican, became one of the most forceful spokespeople on climate and security. In his early career, Senator Warner was secretary of the navy, and he served for decades on the Senate Armed Services Committee, as both chair and ranking member. I first got to know him in 1987 when I joined the staff of the Senate Armed Services Committee. Environment and climate change were not our typical topics of discussion either during my service on the committee or during my tenure at DOD. Patrician and gracious, Senator Warner was often known as the senator who had been briefly married to the glamorous actress Elizabeth Taylor. In his fifth and last term, Warner chaired the Senate Armed Services Committee from 2003 to 2007, hearing directly from members of the CNA Military Advisory Board. Many of the generals and admirals on the board were his close professional colleagues, such as Admiral Joe Lopez and General Sullivan, and suffice to say, their climate analysis made a big impact on him.

Not only did Senator Warner cosponsor the landmark pieces of climate legislation mentioned earlier; after his retirement from the Senate in early 2009, he devoted much of the next decade to advocating for climate security, frequently flying around the country with the generals and admirals of the CNA Military Advisory Board to speak at local venues. In testimony before Senator James Inhofe, his former colleague and a climate skeptic, in 2009, he stated the following, paraphrasing the CNA Military Advisory Board's first report:

> Over my thirty years in the U.S. Senate working with military men and women and their families, I left convinced that, if left unchecked, global warming could increase instability and lead to conflict in already fragile regions of the world. We are talking about energy insecurity, water and food shortages, and climate driven social instability. We ignore these threats at the peril of our national security and at great risk to those in uniform who must operate, on orders of our President, the sea

lifts, the air lifts, and other missions to alleviate humanitarian suffering or sovereign instability in remote regions of the world.[52]

While our CNA report did not heal the partisan wounds that would continue to stall action on climate change, it would lay the groundwork for the field of climate security and, consequently, for climate change becoming widely accepted throughout the US and international security community. In the following chapters, we will explore how that community is facing the multiplying threats posed by increasingly turbulent conditions around the world.

CHAPTER 4

# *Melting Ice and Rising Tensions in the Arctic*

THE US NAVY HAS BEEN TRACKING Arctic conditions even longer than climate scientists have. Since the dawn of the nuclear age, both the United States and Russia have guarded their homelands with nuclear-powered submarines deployed in Arctic waters. In fact, the Russians consider the High North a "bastion" for their ballistic missile submarines of the fabled Northern Fleet. These submarines are on regular patrol and have long played a *Hunt for Red October*–style cat and mouse game of stealth and secrecy. Defense experts count on nuclear submarines to be the most secure leg of the nuclear triad (land, air, sea) because they are the hardest for enemy aircraft or missiles to target.

Admiral James Foggo is a nuclear submariner whose career has, like mine, spanned the Cold War to the climate era. After graduating from the US Naval Academy at Annapolis, Maryland, in 1981, Jamie Foggo found himself aboard the submarine USS *Sea Devil* in the Arctic in 1985. After weeks below water, the crew was scheduled to get some air and a break from the murky depths, surfacing near the geographic North Pole. This was in an era before GPS and with limited communications, and

Lieutenant Foggo, age twenty-six, was looking for the right thickness of ice to enable the sub to break through. Foggo struggled to find the optimal ice conditions and circumnavigated the North Pole numerous times looking for them.

The following morning, while Foggo was having coffee, the captain issued an order for the "ice pick" maneuver—meaning it was time to surface. Foggo ran back to the control room to help the captain with the detailed operating procedures necessary to surface a submarine in the ice. The boat must be positioned properly, with all the masts and antennas angled to avoid damage. In the first attempt, the ice was thicker than predicted and they didn't blast through, costing use of precious air in ballast tanks. But the second try was a charm, the sub emerged from the ice, and the crew spent twenty-four hours in the fresh air, performing such Arctic antics as donning a Santa Claus suit to hit golf balls. "You can't imagine the morale of the crew to have an opportunity to say that you've done this," Foggo told me.[1]

Fast-forward to 2001, sixteen years later, when Foggo was aboard another nuclear submarine, the USS *Oklahoma City*, north of the Arctic Circle. He had better instruments than in 1985 to locate where to surface, but when he found a good spot to do so, there was not enough ice to allow his sailors to leave the boat safely. He recalled an "azure blue sea" at that site, and where he expected to find ice, he found plastic trash instead. In just a few short years, Admiral Foggo, who would become the four-star commander of US Naval Forces Europe-Africa, had witnessed how climate change was opening the Arctic. He didn't yet know, however, how fast the sea ice would retreat and how much this would contribute, along with rising revanchism in Russia, to a new era of tension in this region.

The CNA Military Advisory Board sounded alarm bells about these changes in 2007, warning, "The Arctic, often considered to be the proverbial 'canary' in the earth climate system, is showing clear signs of

Figure 4-1. Submarine surfacing through ice (photo credit: US Navy)

stress." We noted that a 2001 US Navy study concluded that an ice-free Arctic would require an "increased scope of naval operations."[2]

The fact is that the Arctic is now warming at least three times faster than the rest of the globe, and its unique ecosystems are among the world's most vulnerable to climate change. This warming is leading to new extremes in sea and air temperatures, retreating sea ice, dramatic permafrost thaw, rapid ocean acidification, and complex biodiversity changes. Indigenous communities, who have stewarded the Arctic environment from time immemorial, are on the front lines of these changes. Beyond the direct environmental impacts, the loss of ice means ships can more easily navigate the Arctic Ocean and people can more easily extract oil, gas, and minerals—at least in the medium term. Greater access to these natural resources is leading to a new kind of arms race as countries compete to see who can control the top of the world.

Russia, in particular, is eager to monetize its many assets in the Arctic, including oil and gas shipping lanes. Its military buildup in the region increases the likelihood not only of conflict but also of accidents, such as fuel leaks, at sea. None of this is good for our armed forces, who must navigate growing threats to coastal military installations and increased maritime traffic—in increasingly stormy seas—which could lead to greater demand for search and rescue operations and natural disaster responses.[3]

Admiral Foggo was a direct witness to this Arctic opening from 1985 to 2001 and has been among those supporting more capability for the US Navy to operate in the region. He is not the only one.

## The Navy Recognizes Climate Change in the Arctic

Rear Admiral David Titley was the navy's oceanographer when, in 2012, Admiral Gary Roughead, then chief of naval operations—the navy's top admiral—asked him to create a task force on climate change. Dave Titley, a slender man with an intense focus, is fond of a common expression in the military: "What my boss thinks is important, I find fascinating." Titley was certainly fascinated by climate. He had spent his entire career studying how weather and sea conditions affect naval operations, whether amphibious landings or enforcement of no-fly zones by fighter jets.

But he was agnostic about climate *change* until leading the navy's task force on the subject. When he looked at the evidence, he saw how conditions were becoming harder to predict and were threatening key operations, especially in the Arctic. During the Cold War, when Titley came of age, he could rely on sea ice being a certain thickness at a particular time of year based on historical record, allowing him to plan submarine patrols. Now, with the Arctic warming and ice disappearing, the historical record can no longer be used to predict the future.

Indeed, since the National Oceanic and Atmospheric Administration (NOAA) began releasing an annual "Arctic Report Card" in 2006, every year has shown less ice at the top of the world.[4] Along with retreating sea ice, climate change is leading to warmer ocean waters and melting permafrost—the frozen soil that covers two-thirds of Russia and one-quarter of the landmass of the Northern Hemisphere.[5] Thawing permafrost could put as much as 50 percent of the Arctic infrastructure built on it at high risk of damage by 2050, requiring tens of billions of dollars in maintenance and repairs and leading to the release of significant amounts of methane, a potent greenhouse gas.[6]

The navy has witnessed this warming firsthand. Since 1958, it has conducted an annual exercise called ICEX (Ice Exercise) to understand how submarines surface in the ice. This exercise began at the height of the Cold War, when the USS *Nautilus* broke through the polar ice.[7] By the early 2000s, however, the navy needed to learn more than how submarines can safely surface. It began using ICEX to measure and monitor changing ice conditions in the Arctic. In 2016, ICEX took place over a five-week period and included two hundred participants from four nations: the United States, Canada, the United Kingdom, and Norway. By 2020, those five weeks had been cut to three because the ice was no longer strong enough to support Ice Camp Sargo.[8]

What do these changing conditions mean for naval operations in the future? As Titley said in congressional testimony in 2011, "We are witnessing a failure of imagination" when it comes to how climate change affects military readiness.[9] His sentiment was echoed in the CNA report by Admiral Skip Bowman: "Our nuclear submarines operate in an unforgiving environment. Our Navy has recognized this environment and has mitigated the risk. . . . We should begin planning for a similar approach in dealing with potential climate change effects on our national security."[10]

## A Gray Zone atop the World

In 2015, Dave Titley and I visited Svalbard, a Norwegian archipelago in the Arctic. Svalbard has long been a strategic outpost for the West, located at a critical geographic juncture close to both the North Pole and

Russia. Our Norwegian hosts wanted us to see how both climate change and Russian activities were altering the neighborhood.

As we flew from Oslo to Svalbard, the sky brightened with the increasing sun of the spring sky. The snow twinkled on small hills surrounding the even smaller settlement. The multicolored buildings and well-kept streets could have been in a village in the Alps. Yet here we were, almost at the ends of the earth, 78 degrees North. Once a remote Arctic island with just a few hundred residents, Svalbard has become a popular tourist spot for Norwegians and others. In early April, the sun barely sets on the horizon, and the sky turns a deep rose around midnight, with the light returning a few hours later.

Our Norwegian hosts took us to Svalbard's highest point, where a satellite station tracks global communications. Driving from the picturesque town up the mountain, we saw little white domes popping out of the snowy point, as if arching their backs to the sky. These domes serve as crucial downlinks for data from polar satellites passing overhead. The data they collect is vital for both science and security, from images of shrinking glaciers and forest fires to eroding coastlines and lakes gone dry.

Svalbard's position atop the world gives it a front-row seat to Russian Arctic operations, including the submarines that slide toward US waters. Ships from Russia's Northern Fleet routinely pass through the so-called Bear Island Gap between mainland Norway and Svalbard in their silent trek from the shallow Barents Sea to the deeper North Atlantic. There, the subs can deter other nations from using the waters or disrupt deep-sea communication cables or trade routes between ports. These stealthy operations are being complicated by warming, and increasingly navigable, waters. In 2020, for instance, the ice at the North Pole was too thin for Russia to test its newest icebreaker.[11]

Russia has been increasing its activity around Svalbard in recent years. The 1920 Svalbard Treaty gives Norway sovereignty over the territory but allows peaceful economic and scientific use by other nations, including Russia and China, both of which have research stations there. "Warlike purposes" are prohibited by the treaty, and while that term is undefined, Russia now regularly invokes it to protest visits to Svalbard by the Royal Norwegian Navy.[12] Meanwhile, Russia also seeks to ensure its own presence on Svalbard while raising tensions with the West.

These "gray-zone tactics," similar to those used in Crimea and other neighboring states, do not always look like armed conflict. Indeed, they can appear quite benign, from promoting Russian tourism in the area to conducting research. As one scholar noted, "The real push from Moscow when it comes to Svalbard will be occurring onshore, in plain sight, well within the bounds of the Treaty."[13] These campaigns, whether subtly camouflaged "little green men" in Crimea or Russian researchers on Svalbard, could be the trip wire for more mischievous activities. The changing climate itself could become the justification for further foreign intervention on Svalbard, under the guise of needing more scientific presence to monitor melting and warming conditions.[14]

In 2021, the first National Intelligence Estimate on Climate Change warned the United States' top policymakers that "Arctic and non-Arctic states almost certainly will increase their competitive activities as the region becomes more accessible because of warming temperatures and reduced ice. . . . Contested economic and military activities will increase the risk of miscalculation."[15] What that really means is that there is an accident waiting to happen in the Arctic and we are not prepared.

What might that accident look like? Let's look now to the other side of the Arctic, where Alaska and Russia are less than sixty miles apart in the narrow Bering Strait. Alaska senator Lisa Murkowski, warily eyeing the Russian military buildup in Ukraine in early 2021, noted, "Right now, the eyes of the country are trained on the Ukraine border down there. But who's to say that while everybody's looking that way, we don't see other activity on the other side of the country—the side of the country that's 57 miles separated by water from the United States?"[16]

## An Arctic Accident Waiting to Happen

One of the military's jobs is to imagine the types of situations Murkowski alluded to. These scenarios, sometimes called war games, are designed to prepare defense and intelligence analysts and practitioners for various conditions on the battlefield. Climate change has produced a different "battlefield" for just about every scenario a planner can now imagine, altering the very physical foundations of the geostrategic landscape.

In 2018, as part of a symposium called Arctic Futures 2050, I co-organized and "played" a future Arctic accident with US national security professionals and researchers from the Polar Institute of the Woodrow Wilson International Center for Scholars, the US Navy, the US Coast Guard, the Center for Climate and Security, and leading nuclear and climate scientists from Sandia National Laboratories.[17] The premise was this: Imagine a Russian nuclear icebreaker navigating the narrow Bering Strait, where the United States and Russia are only thirty miles apart at the strait's narrowest point. The Russian ship is escorting a Chinese vessel carrying liquefied natural gas (LNG) from Russia's Arctic energy capital, Yamal, to Shanghai. With storms having become less predictable in recent years, an unusually strong storm moves through the strait, pushing the Chinese vessel against an uncharted object, and fuel begins to leak.

Our task was to consider possible consequences of the leak and how decision-makers would react. Would these contaminants reach American shores? Could they affect Alaskan fishing grounds, which could be even more important several decades from now as waters warm and fish stocks move north? What about the Chinese and Russian sailors whose lives may be at risk? Who would lead the search and rescue efforts? Would the Russian authorities be truthful about what they do and do not know about the condition of their icebreaker and its nuclear reactor? Or would they cover up the facts, as they have been known to do in previous nuclear accidents, from Chernobyl to the *Kursk* submarine disaster? These questions, and more, are not just rhetorical. They will confront and challenge our security and defense forces as we navigate this unprecedented territory.

## Tipping Points in Arctic Climate and Security

Russia is the biggest Arctic power, any way you measure it. It has the longest Arctic coastline, and approximately 20 percent of its gross domestic product is derived from activities in its Arctic territory, mostly from fossil fuel extraction—a practice that is among the primary causes of accelerated climate change. Russia, under President Vladimir Putin, is determined to chart an Arctic future that takes advantage of Russia's geography.

With the country's Arctic area holding 14 percent of Russia's oil and 40 percent of its gas reserves, Putin's Russia intends to remain the global gas station for as long as possible. And the urgency has arguably only increased since Putin's war on Ukraine has accelerated the ambitions of Western nations to decarbonize their economies to free themselves from the tethers of Russia. Russia's logic is, essentially, to ramp up extraction and bring these fuels to market before they become a stranded asset in the decarbonization era. Even as many countries move to wean themselves off Russian fossil fuels, Russia will be selling to willing buyers in Africa, Asia, and Latin America.

Putin also dreams of converting the Northern Sea Route, which hugs the Russian coastline, into a toll road for ships transporting goods from ports in Asia to Europe.[18] And China has laid out plans for a Polar Silk Road connecting Shanghai with Rotterdam.[19] In the medium term, the shipping time for these routes could be reduced by as much as 40 percent as a result of melting ice, according to the National Intelligence Estimate. That alone would be a big prize. Although Russia and China see a melting Arctic as a transit opportunity, the unpredictability of the climate also creates complications. In 2021, numerous cargo ships from various nations became stranded along the Northern Sea Route when winter sea ice set in earlier than expected.

Yet Putin remains undeterred in his plans to further develop the Arctic—even by the consequences of his unprovoked war in Ukraine. Since economic sanctions by Western and allied nations have reduced options for Russia's development of the Northern Sea Route and upended Russia's traditional oil routes to Europe, Putin is now scrambling to become the supplier of choice to Asian markets. Russia is also making risky decisions to pursue its Arctic development projects in spite of economic sanctions. In 2023, Russia's state nuclear energy company, Rosatom, announced plans to use non–ice class oil tankers along the Northern Sea Route to deliver Arctic crude oil to Asia. Shipping oil in non–ice class tankers risks a major oil spill in the Arctic, where cleanup is difficult at best and impossible at worst. Such an accident could make the *Exxon Valdez* cleanup look easy. Admiral Thad Allen, former commandant of the US Coast Guard, who served during the Deepwater Horizon oil spill in 2010, told me that "an oil spill in the Arctic creates another level of complexity based on limited access, lack of infrastructure, and multiple legal regimes. . . . We must pay attention to these risks and do more to mitigate them."[20]

These actions are keeping Russia's Arctic neighbors up at night, including the commander of the Royal Canadian Navy, Vice Admiral Angus Topshee. I shared a virtual stage with Topshee in March 2023 to discuss environmental security in the region. He explained that in response to retreating sea ice, rising temperatures, and increasingly aggressive activity by Russia, China, and others, the navy had upgraded its Arctic patrol vessels since 2020.[21] Topshee noted that the Royal Canadian Navy had recently sailed through Canada's Northwest Passage for the first time since 1954, almost seventy years ago.[22] From his vantage point, with the responsibility to patrol for illegal fishing in all Canadian waters, including in the Canadian Arctic, he expressed concern that the Central Arctic Ocean Fisheries agreement (Agreement to Prevent Unregulated High Seas Fisheries in the Central Arctic Ocean), which became effective for sixteen years starting in 2021, might not last.[23] He was worried that with fish stocks moving north due to warmer temperatures, countries like China and Russia would be tempted to fish illegally in the Central Arctic Ocean.[24] I share his concern.

Sometimes I think China's primary Arctic ambition is fishing. In the summer months, off coastal Alaska, Chinese vessels outnumber American ones in these increasingly productive waters.[25] In 2018, China released its first formal Arctic policy, declaring itself a "near-Arctic state," a term the Arctic coastal states, such as the United States, have dismissed. And when China signed the Central Arctic Ocean Fisheries Agreement, it received a major benefit conferred on non-Arctic nations: the opportunity to conduct "research and observation" just outside the Central Arctic Ocean. It seems all too likely that China is using this research opportunity to determine when and how the vast Central Arctic region could have a viable fishery. With China's current record of illegal and unreported fishing activities around the world, especially off the coasts of South America and West Africa, its signing of the agreement may serve as convenient cover for its future fishing plans for the Central Arctic.[26]

In these vast and increasingly open waters, the space between public statements and reality can be wide. Disinformation about climate change and climate action is already a growing concern even outside the Arctic. Not only does Russia flood social media with disinformation about the war in Ukraine, but a Russian lawmaker in 2021 also claimed the United States used a climate weapon to prompt an unusually warm winter in Russia.[27] The term "climate weapon" refers to the prospect of raising or lowering temperatures or altering precipitation through geoengineering techniques, a growing arena of climate intervention research and development.

Since Putin's invasion of Ukraine in February 2022, cooperation between the United States and Russia on Arctic matters has been almost nonexistent. For instance, the Arctic Seven halted their participation in the Arctic Council when Russia was chair, and the only activities that resumed at working-group level excluded Russia.[28] Norway's chairmanship, which started in June 2023, might open new opportunities, although resumption of cooperation with Russia seems unlikely. US–Russian military collaboration in the Arctic—including through the Arctic Security Forces Roundtable—was halted even earlier, when Russia invaded Crimea in 2014. The channels of effective communication have dwindled.

In sum, dramatic and likely irreversible climate change in the Arctic, in the form of sea-ice retreat, rising temperatures, and permafrost thaw, combined with preexisting concerns about Russia's Arctic ambitions and rising tensions elsewhere in the world, has led to a new security paradigm. The United States must now defend its Arctic border from risks more traditionally associated with open water in the rest of America: illegal fishing, human trafficking, oil spills, and tourist accidents.[29] Of equal concern is the potential global rush for resources across the Arctic, from hydrocarbons to critical minerals in the deep-sea bed. Expected to become more accessible as a result of retreating ice and the development of new technologies, the availability of these resources poses a greater risk of environmental damage.

**Drinking with the Russian Bear**

Hard as it is to imagine today, there was a moment of détente, and even cooperation on environmental issues, with Russia in the Arctic. During my time at the Pentagon in the 1990s, I was working to clean up the remnants of the arms race, not just in the United States but also near the North Pole. And Russian officials were my unlikely partners.

It all began with an unusual trip to my office by the secretary of defense at the time, William Perry. If the secretary of defense was coming to me, to my office, I knew it was serious—and I was right. Perry had just met with his counterpart in Norway, and our ally needed help. Back then, the Norwegians were less concerned about the Russian nuclear-powered submarines sliding past Svalbard and more concerned with the decommissioned ones rotting on the seafloor and contaminating their waters.

Just sixty miles from the border with Northern Norway, the Russian Arctic port of Murmansk and surrounding military bases are home to the Soviet nuclear navy, part of the fabled Northern Fleet, as well as its highly radioactive spent fuel.[30] When the Cold War ended in 1990, many of the Russian nuclear submarines were left to sink at the piers in Murmansk. In 1993, the Russian government admitted to dumping radioactive waste and other poisons in the ocean and confessed that Soviet submarine accidents had also released toxic material.

Norway's defense minister, Jørgen Kosmo, worried that these sunken subs were a "slow-motion Chernobyl" at sea.[31] And he wanted to reduce Norwegians' concerns about contamination of prized fishing areas in their Arctic waters, which feed much of the country and are a major source of exports. Indeed, the Barents Sea has the richest cod fishery in the world and is also an important habitat for haddock, red king crab, walruses, whales, polar bears, and many others. Who would buy Norwegian fish if they were thought to be poisoned by Russia's nuclear waste?

Working with key experts in the US Navy, Army, and Air Force and my own staff, we put together a plan to engage the Russian military and develop projects that would remove liquid waste from the sunken Soviet submarines.[32] The signature project was a large metal cask that would store the radioactive material and allow for safer shipment to inland storage. But first, we had to get an agreement—the Arctic Military Environmental Cooperation, or AMEC—signed by the three secretaries or ministers of defense of Norway, Russia, and the United States. They

Figure 4-2. AMEC cask for liquid waste from spent nuclear fuel, St. Petersburg, Russia, 2000 (photo credit: author's personal collection)

would meet at a summit of the North Atlantic Treaty Organization in Bergen, Norway, in September 1996.

During those halcyon days after the breakup of the Soviet Union, inviting a Russian defense minister to a NATO meeting was not only possible but part of normalizing relations with the new Russia. However, getting the Russians to sign an agreement was no easy task. Even though the United States and Norway were offering to provide the money and much of the engineering and technical support, the Russians had their own bureaucracy.

They also had their own traditions. From my very first visit to Moscow to negotiate the AMEC agreement, every day ended with an evening of toasts. Since I was usually the most senior member of the US delegation, I was expected to drink the toasts the Russians made in my honor and reciprocate with my own. After each speech, we had to quickly down a shot of vodka. I learned the hard way that I could not possibly keep up with my Russian hosts. I could barely move the morning after a vodka-and-caviar-filled evening with Russian generals.

That's when I created the "designated drinker." On every subsequent trip I made to Russia in the 1990s, American generals and admirals were part of the delegation. I quickly learned that military officers not only could hold their vodka better than I could but also were much more eager to do so! And, thankfully, being pregnant at some of these meetings worked in my favor—the Russians did not think I was copping out. To the contrary, they were fond of children and designated my young daughter, Natalie, the "mascot" of our program.

The negotiations dragged on through 1995 and into 1996. With the NATO meeting looming, we inked an agreement in August, just a month before our deadline. (Our invitation for the Russians to join us at an environmental conference on the Hawaiian island of Oahu may have helped sweeten the deal.) On September 26, 1996, the three defense leaders formally signed the AMEC Declaration, highlighting "the need

Figure 4-3. US and Russian delegation at the A. F. Mozhaysky Military-Space Academy in St. Petersburg, May 14, 1997 (photo credit: author's personal collection)

to ensure the conservation and sustainable use of the Arctic environment" and noting the military's role in the "established framework for international environmental cooperation."[33]

Over the next five years, I regularly traveled to Moscow to keep up the pace of projects we had jointly developed with Russia and Norway. The AMEC program formally ran for a decade. The landmark agreement not only reduced toxic pollution and headed off an economic calamity for an important NATO ally; it also accelerated the dismantling of the Russian nuclear submarine fleet. Furthermore, it showed that cooperation with Russia was possible.

Remarkably, AMEC was one of several joint environmental security efforts between the United States and Russia during the 1990s. As part of a 1995 bilateral agreement between the US Department of Defense and the Ministry of Defence of the Russian Federation, I delivered lectures on US military environmental education and training at Russian military academies in Moscow and St. Petersburg. We also traded maps that depicted environmental conditions at specific US and Russian military bases: the United States produced a map from satellite imagery of Russia's base at Yeysk, and Russia gave us a map of Eglin Air Force Base in Florida.[34] Russia also produced a map documenting possible changes to the coral reef at Johnston Atoll in the Pacific Ocean. This exchange of previously classified imagery was one of the high points of post–Cold War cooperation.

Perhaps most remarkably, I coauthored an article with my counterpart in the Russian ministry of defense, Lieutenant General Sergei Grigorov of the Directorate of Ecology. We observed that "environmental degradation and related conditions may contribute significantly to instability around the world, which in turn would threaten our national interests in regions of strategic importance. Military environmental cooperation can also help promote democracy and trust, and increase the capacity to address environmental problems."[35]

Unfortunately, those days of military collaboration are now long gone. By the early 2000s, relations between the United States and Russia would rapidly deteriorate as Vladimir Putin rose to power. America's focus would shift from helping rid Russia of nuclear weapons to preparing for an Arctic without ice. From the era when Soviet and American subs poked up from the ice to peer at each other, we now wonder how long there will be enough ice for these cat and mouse games.

The spirit of post–Cold War cooperation that enabled the United States and Russia to work together on common environmental concerns in the 1990s, from AMEC to the sharing of formerly classified intelligence images of military sites, has given way to rising tensions over Putin's unprovoked war in Ukraine and aggressive behavior in the Arctic, with potential spillover to other Arctic nations. The rosiest outlook for this region is that increasing global attention will create pressure to reduce these environmental security threats, at least in the form of information sharing, scientific exchange, and environmental cooperation—similar to what occurred throughout the Cold War. Although that is a possibility, we need to plan for a more dangerous—and more likely—future.

CHAPTER 5

# *Drought, Oil, and Power in Africa and the Middle East*

When Robert Hayward deployed to Iraq in 2005, he went kicking and screaming, but not for the reasons you might imagine—in fact, quite the opposite. As a pilot of the highly coveted Blackhawk helicopter and a self-described "hangar rat," Hayward wanted nothing more than to fly military missions. During flight school, he spent almost all his time at the aircraft hangar, learning anything and everything there was to know about US Army aviation. He graduated at the top of his class, and soon after, he led the flight detachment for President George W. Bush at his ranch in Waco, Texas.

But Hayward wasn't sent to Iraq to be a pilot, and he was none too excited with what appeared, at first, to be a mundane mission. His unit was charged with helping to stabilize Baghdad after the US invasion by supporting the Iraqi military and ensuring the city had working power and sanitation. The army needed Hayward to restart the city rather than fly helicopters. "First order of business," he said, "was to turn the lights back on."[1] So Hayward's team worked directly with civil engineers in Baghdad to repair critical infrastructure across the city, including electricity, sewage, water, and oil and gas pipelines. From the outset, Hayward knew that all logistics operations relied on dependable sources of energy, but he would quickly discover that these were also critical to peace in the area.

Just outside the military base, Hayward explained, the environment in Baghdad was unstable, hostile. Residents were suffering. People were hungry, thirsty, without essentials for survival. While Hayward's unit was there to establish bases and ensure the Iraqi army's success, all the resources and expertise provided to the Iraqi soldiers were needed just as desperately by civilians. Hayward realized that he and his team needed to look beyond their military purview and broaden their mission, so they expanded operations and began sharing energy generators and water purification systems with the local population. "Our generators became important. They were viewed as olive branches to those local communities," Hayward said.[2]

As the Iraqi people gained access to critical infrastructure, from housing to electricity, Hayward noticed that hostilities were cooling. "We were alleviating stressors," he said, and, in parallel, experiencing an ebb in local unrest. It is perhaps an obvious point but one that is too often overlooked: peace requires meeting people's basic human needs, including food, water, and—in today's society—energy. When more Baghdad residents had access to these essentials, they could turn their efforts to rebuilding.

## Multiplying Threats

Hayward's experience is a perfect example of the direct correlation between stability and access to natural resources—resources that are disappearing as climate change stresses already fragile regions around the world. Nowhere is this connection more evident than in the Middle East and North Africa, areas with a naturally arid climate and a fragmented and volatile sociopolitical landscape. As climate change further strains these environments, threats of violence are growing—even in countries with the most stable governments.

There's a poignant irony to the region's geography: it is bone dry, yet its deep oil and gas reserves power much of the planet. These reservoirs and the narrow straits they travel—the Suez Canal and Strait of Hormuz—are so vital to our global economy that the industrialized world currently could barely operate without them. The world's need for energy creates a vicious cycle as the burning of fossil fuels contributes to climate change and the region increasingly becomes a desert. The Middle East and North Africa have seen temperatures soar, winter rains dry up, and flash floods increase, threatening the livelihoods of farmers and contributing to hunger, thirst, and heat-related deaths for its residents.[3] As incomes dwindle, more will fall into poverty, exacerbating existing inequalities and grievances, along with governance problems, and making the region more vulnerable to extremism and violence.[4]

That is particularly dangerous in a region that has been perennially beset by conflict as various factions struggle for money, power, and land. For complex historical and geopolitical reasons, some nations, such as Yemen and Somalia, still barely function, while others, from Israel to the United Arab Emirates and Saudi Arabia, have had highly productive economies despite ongoing security woes. Yet even the wealthiest countries will have to adjust as the world transitions away from fossil fuels, as almost two hundred nations agreed to do at the 2023 United Nations Climate Change Conference (COP28) in Dubai, United Arab Emirates.[5]

From this perspective, the region's current challenge is to turn the corner toward clean energy while still building prosperity and political stability and effectively responding to climate impacts. This goal is further complicated by ongoing terrorism and conflict, including the dramatic setbacks to peace and stability as a result of the Hamas terror attacks and the Israeli response that began in October 2023. Today, the region's major energy producers, such as Saudi Arabia, have seen economic growth as Russian oil exports have declined as a result of the war in Ukraine. If they succeed in their energy transition, these nations are poised to become even more strategically central to global geopolitics because they are among the lower-cost and lower-emissions oil producers.[6]

These changing dynamics present serious questions for both regional leaders and the world. How will the transition to renewables affect the region's role in the global economy? As water disappears, heat waves intensify, and desertification envelops more and more land, will local governments respond effectively? When heat waves, thirst, and hunger force migrants to flee their homelands toward urban areas, will there be adequate policy responses to protect their human rights and address their basic needs? Or will extremist groups and terrorist organizations take advantage of these vulnerable populations?

For the US military, the threat of further destabilization in the region is a pressing concern. General Anthony Zinni has witnessed the dangers of instability firsthand over the course of a thirty-five-year military career, leading relief efforts in the former Soviet Republics, Somalia, Turkey, and Iraq. Throughout these operations, he grew to understand that the depletion of natural resources has dire effects on the security of local communities: "If the government there is not able to cope with the effects, and if other institutions are unable to cope, then you can be faced with a collapsing state. And these end up as breeding grounds for instability, for insurgencies, for warlords. You start to see real extremism," he reflected years later.[7]

In 1995, General Zinni served as commander of the Combined Task Force for Operation United Shield, securing the withdrawal of United Nations forces from Somalia following humanitarian efforts in the nation. During this time, he met countless farmers whose fields had been decimated by drought. Unable to grow crops as they once did, these farmers turned to the sea for food and income. "They weren't fishermen.

They were farmers, but they had lost their land and went after the nearest source of food," General Zinni said.[8]

Despite living along the water, these communities had previously left their far-reaching coastline largely untouched because fish were not a traditional part of their diet. But as the prolonged drought and corresponding famine decimated the Somalian people, they started demolishing their own coral reefs—"really beautiful reefs," General Zinni said—in order to gain better access to the fishing grounds. Witnessing people who had spent their lives cultivating the earth now forced to destroy their environment in order to survive was a raw experience for General Zinni and led him to advocate for the integration of climate change into military operations.

As he explained, environmental issues and national security are so inextricably linked that they can no longer be considered in isolation. "Sometimes we may think you sort of have a silo that's military and you stay in that lane," General Zinni said. "But you have to understand how these [climate and environmental risks] can make those situations, military situations, worse."[9]

In turn, military operations can worsen environmental problems. Throughout history, wars have been fought for national interests, to preserve or acquire territory, but certainly not to protect the environment, which is often a casualty of conflict. In the cases of Iraq and Afghanistan, the US military was focused on fighting terrorism and sometimes lost sight of the environmental damage to the region, General Zinni said.[10] In Iraq, for example, clearing natural vegetation for military action worsened sandstorms and dust storms, forcing residents to leave rural areas and threatening the health of those who stayed. The scars left by those operations remain long after the fighting subsided.

General Zinni's realizations reflect a broader evolution in the military's thinking about our role in creating and responding to environmental threats—a change that is happening because uniformed military leaders are seeing the consequences on the ground. Few embody this dramatic shift in thinking better than General Chuck Wald, who, as special assistant to the US Air Force chief of staff in 1997, was a key proponent of a military exemption from the Kyoto Protocol. During the negotiations, General Wald argued vehemently that capping the air force's greenhouse gas emissions would hurt military preparedness and threaten mission success. Back then, he was my bureaucratic opponent, contending that global climate protections did not and should not influence the military. Ten years later, we were working side by side on the CNA Military Advisory Board to address the national security implications of a changing climate.

What changed from 1997 to 2007? General Wald witnessed, for himself, floods and droughts that caused people to go hungry and thirsty, spurring violence and unrest and threatening regional stability across Africa. In his final military assignment, General Wald was in charge of all US forces in Africa. Prior to the establishment of a separate US Africa Command (US AFRICOM) in 2007, US operations in the African continent were coordinated by the US European Command's deputy commander, a role General Wald held from 2003 to 2006.

When he first assumed this command, General Wald discovered that US military plans had not been updated since the conclusion of the Cold War, when the focus was on fighting a land war in Europe against the Soviet-controlled Warsaw Pact. Why this extensive delay and seeming oversight? No one was sure of the key threats facing the region after the Soviet Union collapsed. So General Wald set out to update US military missions and capabilities for Africa, traveling the continent as he developed his expertise and often providing food, water, and medical supplies as he went. Through his travels across the continent, General Wald saw directly how the changing climate was threatening security in areas US forces were trying to stabilize. And he realized it would be vital to consider climate change when revising operational plans.

While climate risks are only now, in the past few decades, becoming part of defense decision-making, weather has always been considered during military planning. As General Tom Middendorp, former Dutch chief of defense and my co-lead at the International Military Council on Climate and Security, explained, military service members learn very early on that every operation begins with an analysis of weather and terrain.[11] But until recently, the military didn't necessarily connect that variability to climate change. For General Middendorp, the link became obvious during his second deployment in Afghanistan in 2009 while leading a multinational task force in the Uruzgan Province. There, he saw that the troops were fighting symptoms of a deeper problem—one rooted in the damage climate change was inflicting on local communities.

In a village called Chorah in the southern part of the province, General Middendorp and his task force fought fiercely for several days to clear the Taliban from the area. Although they finally succeeded, tensions in the village remained high and the Taliban remained able to return at any moment. "It took a while before we found out the root cause of those tensions was scarcity of water," he said.[12] The village farmers desperately needed water but could not agree on a fair way to share it. The Dutch forces brought in negotiators to craft a water management plan for the village, equitably dividing the scarce resource. Once this deal was created, the tensions subsided.

"The Taliban couldn't return anymore because the Taliban had been leveraging the tensions in that village," General Middendorp said. "They were brokering solutions on water and division of water, which gave them a power base in that village. Once we had a solution in place, they didn't have that power anymore."[13] After the Dutch military's work to secure a fair water management plan, Chorah became the most stable village in the province.

General Middendorp's experiences in Somalia, Iraq, and Mali led him to similar conclusions that climate change is acting as a threat multiplier, making resources even scarcer as populations bulge—particularly in countries such as Mali, which has one of the highest birth rates in the world.[14] This combination of population growth, poverty, and fragile governance, overlaid with climate change, has led, in General Middendorp's view, to increasing conflict in the parts of Africa and the Middle East most in need of stability.

These pressures, in the midst of a hard political reality, unfortunately meant that water security did not last in Chorah, Afghanistan. Still, that moment of peace and cooperation shows that progress is possible when military leaders factor climate change into their efforts. For Generals Anthony Zinni, Chuck Wald, and Tom Middendorp, there is no other choice. Climate change is threatening the security of this region, and our military must respond if we hope to promote stability.

**Water and Violence**

While climate insecurity manifests in myriad ways, one of the most obvious and serious in the Middle East and Africa is water stress. And just as desertification does not happen all at once, an area does not generally flip from peace to chaos instantaneously. Climate change exacerbates the risk of conflict along a spectrum that begins with civil unrest and localized violence and expands to terrorism, insurgencies, civil war, and, finally, war between nations.[15] At each point in this evolution, water tends to interact with conflict in one of three main ways, according to scholar Marcus King. First, prolonged drought can push an already fragile region toward violent extremism and, in some cases, civil war. Second, water can be weaponized during an ongoing conflict.[16] And finally, in hopeful moments, as General Middendorp saw in Afghanistan, water can become a means of cooperation and even peace-building efforts.

A tragic example of the first scenario, prolonged drought amplifying the risk of violence, can be seen in the death and destruction that engulfed Syria starting in 2011.[17] As a semiarid region, Syria is accustomed to dry conditions, but the drought that stretched from 2007 to 2010, including the driest winter on record between 2007 and 2008, devastated the nation.[18] Before the onset of this drought, Syrians were already struggling to get enough water. Local aquifers were overdrawn. Waste from farms was contaminating water. And the failed policies of President Bashar al-Assad's corrupt government were not helping matters.[19]

The drought, made more likely by climate change and stretching for years, helped push the country over the edge, making it impossible for local farmers and herders to secure enough water for their families and livelihoods. Those shortages, along with skyrocketing food costs and widespread disease, led to violence in communities where workers had lived peaceably for generations. As a result, waves of rural workers migrated to Syrian cities, seeking new jobs and the chance for their families' survival. The United Nations estimated that in 2010 alone, approximately fifty thousand Syrian families moved from rural stretches of the nation to its urban centers.[20] There, they were treated as second-class citizens, relegated to makeshift settlements that lacked water and basic sanitation.

In 2011, as the events of the Arab Awakening in Tunisia, Egypt, and Libya unfolded, tensions boiled over in Syria as well. Security forces cracked down on civilian protestors in multiple Syrian towns, leading to conflict among dozens of factions, including the Syrian government, opposition forces, and a collection of extremist groups.[21] As violence engulfed Syria, these groups, including al-Qaeda and ISIS, targeted key water infrastructure and used access to the precious resource to gain power over desperate citizens and recruit new members to join their forces.[22] By 2012, Syria was embroiled in a full-blown civil war, a conflict that would soon bleed beyond its borders and last for five years, with violence continuing to the present day. And it all began with drought and poor water management.

Even when drought does not contribute directly to conflict, in dry, stressed regions water can become a weapon for extremists—a scenario that has played out throughout the Middle East and Africa, from Syria and Iraq to Nigeria and Somalia.[23] While contaminating or blowing up utility plants and dams is unfortunately a common military tactic in war-torn areas, the weaponization of water is a longer-term strategy that involves controlling the resource.

No militant group has used this strategy more effectively or ruthlessly than ISIS. In 2013, it seized control of all seven major dams along the Euphrates and Tigris Rivers, the most valuable water resources in the region. Power over the dams allowed ISIS to expand into Iraq, and the following year the group took over all large Iraqi dams, save one, along these two major rivers. In June 2014, ISIS captured the neighboring Iraqi cities of Mosul and Tikrit, effectively cutting off water from surrounding villages. With control over the area's water, ISIS could claim authority as a governing body and advance its political aims.

But the militants weren't done yet. Just three months later, in August 2014, ISIS seized Iraq's Mosul Dam, which produces the vast majority of the nation's hydropower and is the largest dam in the Middle East. The group unleashed a sixty-six-foot wave of water, destroying downstream towns and cities along the Tigris River. For the United States, this was too much of a threat, and the military launched air strikes the next day. With support from the US Air Force, Iraqi and Kurdish forces spearheaded a counteroffensive attack and reclaimed the dam within ten days.

Yet this setback did not stop ISIS from continuing to weaponize water. In September 2014, the militants cut off Iraqi military forces by diverting water from rivers in the Sirwan Basin in the northeastern province of Diyala. The flooding not only stopped the advance of Iraqi troops but also enveloped nine local villages, harming civilians as much as the military. At times, citizens themselves were the target, as in October of the same year, when ISIS diverted waters from the Khalis tributary of the Tigris River, flooding parts of the Diyala province, including over 780 acres of agricultural land. The militant group subsequently cut off water from this tributary for ten consecutive days, withholding access to drinking water in nearby villages.

While ISIS used water as a literal weapon to flood towns and terrorize residents, its ultimate strategy was to monopolize water utilities, along with gas and electricity. By controlling services typically provided by the government or industry and jacking up the prices, the militant group put an economic stranglehold on the population, coercing civilians to support its cause. Indeed, ISIS's rapid rise to power in Syria and Iraq would not have been possible without its weaponization of water.[24]

Given these horrific stories, it would be easy to view control of water only as a means of violence. But paradoxically, it has also become a catalyst for cooperation and, in some cases, outside broader conflict, a tool for peace building. There is no better example of this than EcoPeace Middle East, a unique nongovernmental organization (NGO) headquartered in Amman, Jordan; Tel Aviv, Israel; and Ramallah, Palestine. Prior to the outbreak of the recent war, EcoPeace worked to bring these historically conflict-ridden groups together to advance water cooperation and confront the shared risks climate change poses for the region.

EcoPeace's work holds a personal, as well as professional, interest for me. Growing up Jewish in America, I took part in the traditional rituals of my religion: Sunday school, synagogue service on the High Holidays, and a Bat Mitzvah. My family celebrated this Bat Mitzvah milestone with a trip to Israel. My primary memories of this visit to my faith's Holy Land are of the immensely dry earth, soldiers in uniform on every Israeli bus, and the incredible capacity of the nation to turn the desert green. With abundant crops on kibbutzim and greenhouses stretching across the arid land, technological innovation was everywhere, from the hills hugging Jerusalem to the biblical Dead Sea. In 2019, I had the opportunity to return to the region, and once again, I was in awe of the

Figure 5-1: Mesopotamian marshes, Basra, Iraq (photo credit: HomoCosmicos)

capabilities of such an arid area. I visited Jordan, Israel, and Palestine with EcoPeace to observe how the respective leaders of this organization were cooperating to share water in this dry land.

This remarkable NGO was founded in 1994 when its cofounder and codirector Gidon Bromberg was studying in Washington, DC, amid the signing of the Israel–Jordan peace treaty and the Oslo Accords. These landmark peace agreements inspired Bromberg. Nevertheless, despite some modest water-sharing arrangements, he was surprised to see that environmental considerations were largely absent from development plans for the region. New hotels, roads, and industrial plants were being planned without consideration of their impact on their surroundings. For fragile areas such as the Dead Sea, the Gulf of Aqaba, and the Jordan Valley, this development could lead to serious environmental damage, which would make peace less sustainable. With ambitions to protect both the land and its people, Bromberg and his Jordanian and Palestinian colleagues formed the first regional NGO to jointly address environmental challenges.[25]

Those challenges are steep. Taps running dry are simply part of life across much of this region, and rising temperatures and prolonged periods of drought are making matters worse. The Israel Meteorological Service projects that average temperatures could increase by 4 degrees Celsius (7.2 degrees Fahrenheit) by the end of the century, with regional rainfall diminishing by as much as 20 percent.

Yet seeds of hope emerged from this scarcity. In 2022, Jordan and Israel agreed to a climate-water-energy cooperation deal, a cross-border agreement inspired by EcoPeace's plan for a Green Blue Deal for the Middle East. Through this new agreement, Jordan was to provide 600 megawatts of solar electricity to Israel, which in turn would supply Jordan with 200 million cubic meters of desalinated water. The plan capitalized on the neighboring nations' comparative advantages, each providing the other with resources the recipient is less able to cultivate. While Jordan's geography makes it difficult to build desalination plants, Israel has long operated five such facilities along its Mediterranean coast. Meanwhile, the small and densely populated Israeli state has limited space for large-scale solar installations, while Jordan is home to vast, sun-drenched desert land that is suitable for massive arrays of solar panels.

"Jordan could become a regional hub for renewable energy, selling renewable energy to the entire region, not only Israel," Jordanian EcoPeace codirector Yana Abu Taleb said. "And imagine all that climate security that we're achieving, all the economic benefits for the country as well."[26]

The second element of EcoPeace's Green Blue Deal is rehabilitation of the biblical Jordan River. The Jordan Valley has become one of the hottest places on the planet, and as temperatures and population ticked up, the river's once rushing waters dwindled to a stream. Today, the flow of the Jordan River is less than one-tenth of average historical levels, which is also the major cause of the shrinking of the Dead Sea (the rest is due to Israeli and Jordanian industries extracting salts and minerals, at a rate of three feet per year, to the point that it has almost become two bodies of water).[27]

Instead of fresh water, Bromberg says, the Jordan River is full of sewage and waste: "That's actually what's kept the river wet—a combination of agricultural runoff, sewage water and saline waters."[28] To start cleaning up the pollution, Egypt, Israel, and Jordan signed an agreement at the 2022 United Nations Climate Change Conference (COP27) in Sharm el-Sheikh, Egypt, based on EcoPeace's work on improving the Jordan River. The deal, however, still needs to attract global investment to improve sewage treatment and help restore the river. That goal is important not just to Israelis and Jordanians but also to the many Syrians who have made their home in Jordan. The rehabilitation of the Jordan River could protect this holy region and its inhabitants from the tragic fate that befell Syria.

For now, as of this writing in late 2023, EcoPeace's admirable work to encourage cooperation and conservation is on hold. Because of the war in Gaza, Jordan has pulled out of the agreed-upon water-for-energy deal, and EcoPeace's cleanup of sewage in the Gaza Strip has been halted.[29] These water security and peace-building efforts will be, at best, postponed—another casualty of the brutal war.

In the longer term, however, hope remains that EcoPeace's efforts will bear fruit. One important component of the organization's work is a host of cross-border educational programs. Drawing together students and young professionals from Israel, Jordan, and Palestine, the courses challenge participants to tackle regional environmental problems without becoming sidelined by politics. Through roundtable discussions, participants learn to listen to the ideas and needs of others while sharing their own thoughts and opinions.[30] Rising above decades, if not centuries, of conflict is not easy, but students recognize that their respective countries cannot solve such massive challenges alone.[31] EcoPeace's programming provides a safe place for conversation and collaboration where students with diverse backgrounds and beliefs can work together for a more peaceful future.

In a region engulfed in violence, with unparalleled strife, intractable politics, and a parched landscape, EcoPeace continues to inspire neighbors to find enduring solutions and points of commonality to counter the forces that divide them.

## Shifting Priorities

As the threats posed by water scarcity, and by climate change more broadly, become harder and harder to ignore, the US military has begun thinking about how to prevent conflicts rather than simply fight in them. That paradigm shift was just beginning as Swathi Veeravalli graduated from the University of Oxford in 2009 with a master's degree in water science policy. "The 'doctrine du jour' is that the military was trying to move away from kinetic operations to focus on stability operations," Veeravalli told me.[32] It was this shift that inspired her to join the US Army Corps of Engineers as an environmental scientist. "I thought I could be part of that pivot towards prevention, towards stopping the business of warfare," she said.[33]

Her goal was inspired by a childhood spent in dry and vulnerable lands. Veeravalli was born in Southern India and raised in Botswana, in Southern Africa. Growing up during a twenty-year drought in the arid nation, Veeravalli was keenly aware of water conservation from a very young age. "We were always very careful with water," she said. "We were very careful with the environment."[34] The experience shaped Veeravalli's career, motivating her to work on the water scarcity and social insecurity issues plaguing Kenya and Sudan.

As she conducted research on water access in these nations, Veeravalli also delved into the nuances of US government funding. The only efforts receiving money at that time were very specific engineering projects, she said, or projects designed to support the warfighter. "Why are we not funding cooperation?" she wondered.[35]

In 2019, Veeravalli joined the US AFRICOM as a foreign affairs specialist. In this role, she was on a mission to show exactly how environmental change was affecting the African continent—and to inspire military planning on the matter. Military leaders might appreciate the link between climate change and security, but what were they supposed to do with that knowledge? The United States unquestionably knows how to respond to natural disasters, but could we reduce these threats to begin with? How could we ensure that the next disaster would not undermine stability in the strategically important regions of Africa?

Veeravalli's answer was simple, if not easy: the United States must develop "mutually assured resilience" with the region. This positive spin on the Cold War term "mutual assured destruction" means investing in critical regions across Africa, Latin America, and Asia, the futures of which are essential for both American and global prosperity. One significant part of that effort is protecting global supply chains, both from shocks to the system and from human rights abuses. Making sure that countries are not caught short of goods in the face of natural disasters or conflicts (or a global pandemic) is key—as is safeguarding workers' safety and human rights.

For instance, the United States' transition to renewable energy currently depends, in part, on mining operations in Africa that supply key minerals for batteries to power electric cars and turbines. China saw the importance of this market years ago and has worked to corner it, offering African nations aid and infrastructure projects in exchange for lucrative mineral contracts. The result is that China largely controls the supply of minerals, while laborers have suffered, with African children being sent into dangerous conditions to mine. If we are to preserve our energy future and Africa's stability, we must counter China's influence and strengthen our relationships in the region. That means being willing to invest in projects that build resilience and reduce poverty and human rights abuses on the continent and elsewhere.

Even with growing support for these investments, there are times when I cannot help but think that peace and prosperity in this region will always remain a distant aspiration, never to be fully achieved—that politics, terrorism, and violence will continue to outweigh peace building and cooperation on water, energy, and climate security. But then I think of the good work that EcoPeace accomplished for more than thirty years, before the start of the most recent war. And I think of the resilience of my ancestors as they fled the Holocaust and earlier wars, hoping against all odds that better times lay ahead.

Today, I find that hope in Veeravalli's vision of mutually assured resilience. In the past, Israelis and Palestinians (at least in the West Bank) have come together over water, as have sworn enemies throughout history. To date, cooperation on environmental issues has not alone been enough to stop the conflict that still rocks the region, yet it remains essential to a lasting peace. It is a seed from which greater collaboration and mutual understanding can grow. If we believe in the adage "Out of crisis comes opportunity," the opportunity has never been greater.

CHAPTER 6

# Navigating Asia's Disaster Alley

LEE GUNN WAS A YOUNG BOY IN 1947, just after World War II, when he sailed with his mother aboard a troop ship, first from Hawaii to Guam and then to the even more remote Micronesian island of Pohnpei. On that tiny atoll, Japan had built a large base before the United States destroyed the nearby Truk Atoll and cut off Japan's forces in 1944. Gunn's father was a lieutenant commander in the US Navy who was helping to rebuild Pohnpei and other Micronesian atolls after the destructive battles of World War II. Gunn recalled that he spent most of ages five to seven with navy construction engineers, known as Seabees, who were charged with fixing the roads and bridges damaged during the war with Japan in the Pacific region. His language got so salty during his time with the sailors that he needed some "correction" when he got to American schools, a few years later.[1]

Gunn grew up in the navy, followed in his father's footsteps, and spent almost forty years in uniform, ultimately becoming the navy's inspector general. His childhood in the Pacific region shaped the way he saw America's role in the world. From a young age, he recognized that the US military's mission went much further than fighting wars. The sailors he knew not only served in combat but also rebuilt bridges and repaired roads to maintain American presence in the Pacific.

Gunn's youth was a lesson in soft power: strategies that use carrots rather than sticks to influence other nations.[2] These operations, such as construction projects, have both humanitarian and defense goals; without any bullets fired, they can shape a region's alliances and political dynamics. Soft power in the form of cooperation on environmental issues is essential to security and is a recurring theme in the military's march toward greater environmental awareness. In chapter 4, for example, we saw how the US military cooperated with Russia in the Arctic both to protect the natural environment and as a means of diplomacy. In the Indo-Pacific region, soft power has evolved from helping communities rebuild in the wake of World War II to assisting them in recovering from the ravages of tropical cyclones and typhoons, worsened by climate change. By the time Gunn was running naval missions, operations included natural disaster rescue and relief, as well as projects to improve water, food, and energy security in vulnerable locations.

In the twilight of his career, Gunn became an ardent spokesperson for military leadership on climate change. Today, when Admiral Gunn thinks about this existential threat, he remembers a plaque on the desk of Vice Admiral Paul Butcher, a gruff, cigar-chomping figure with whom he served in the 1970s: "Lead, follow, or get the hell out of the way."[3]

No navy leader embodies that sentiment better than Admiral Samuel Locklear, whose last assignment was commander of the US Pacific Command. Coming of age a decade later than Lee Gunn, Locklear was a landlocked kid during most of his youth. Born in Macon, Georgia, descended from cotton farmers and mill workers, he moved with his family to rural South Carolina when he was young. He enlisted in the navy at age seventeen "out of frustration" in the waning days of the Vietnam

War. After graduating from a prep school that was a feeder to the US Naval Academy at Annapolis, Maryland, Locklear spent most of the next four decades at sea or otherwise deployed around the world. He told me that he was rarely in the United States, other than "to have my children and make sure my wife's washing machine worked occasionally."[4]

During the course of his forty-three-year career in uniform, Locklear had a front-row seat to a changing world. When he was eight years old, there were three billion people on the planet—and today there are eight billion. On his first deployment to South America in the late 1970s, he said, "We took three ships and we circumnavigated South America. And it was, in retrospect, the end of an era because we had no satellite communication. No GPS. And not many people either."[5] But that rapidly changed. Whether he was sailing into port or flying into a remote region, Locklear was an eyewitness to global growth and with it the awareness, as he said, that "eight billion people are now all clamoring for a piece of the economic pie that the US had successfully garnered in World War II."

Throughout his years in the navy, Admiral Locklear directly experienced the threat that an increasingly unpredictable climate posed to growing populations. When he retired from the navy in 2016, Admiral Locklear joined the Center for Climate and Security. In the foreword to a 2015 report, he stated:

> Today we find ourselves in a period of unprecedented global change—change that is offering many new opportunities, but also introducing significant emerging challenges to the global security environment. Foremost among these emerging challenges are the long-term security implications of climate change, particularly in the vast and vulnerable Asia-Pacific region, where the nexus of humanity and the effects of climate change are expected to be most profound.[6]

Both Sam Locklear's and Lee Gunn's naval careers, from the 1970s onward, exemplify a military on the front lines of climate change and its disruption of lives, nowhere more than in the Pacific.

## Cooperation with Pacific Leaders

When I arrived at the Pentagon in the early 1990s, these officers were about halfway through their careers. I quickly realized that making progress on environmental issues required engaging the combatant commanders stationed around the world. This meant traveling not only close to the Pentagon, such as Norfolk, Virginia, but also to far-flung bases in the Pacific. Hawaii is home to the US Pacific Command (today the US Indo-Pacific Command), headquarters for the greatest concentration of US military forces in the world. Not only does the CINCPAC—commander in chief pacific—have his spacious headquarters on Oahu, but all the US military services—US Army, US Navy, US Air Force, and US Marine Corps—have major divisions and units stationed across the Hawaiian Islands.

One of the first four-stars I worked with closely was Admiral Joseph Prueher, who served as CINCPAC from 1996 to 1999. With a studious and serious demeanor, Admiral Prueher, like most of those with his pedigree (US Naval Academy at Annapolis, navy test pilot), did not suffer fools. If you wanted to get his attention on a matter, you had better know what you were talking about. Back in the 1990s, Admiral Prueher was focused on China's rise; later, when he retired from the navy, he would serve as US ambassador to China. He understood, earlier than many others, that combating growing Chinese influence in the Indo-Pacific region takes more than troops, ships, and aircraft. It also means working with Pacific allies on matters of deep concern to them, some of which are environmental.

For some Pacific Islands nations, the biggest issue was the legacy of nuclear testing performed by the United States, which still required cleanup years after the war. After World War II, the United States forcibly relocated the inhabitants of Bikini Atoll in the Marshall Islands and conducted twenty-three nuclear tests there from 1946 to 1958.[7] Citizens of these islands are still prosecuting health claims against the United States,[8] and this legacy has haunted America's relations in the region to this day.

For other countries in the region, developing professionalism in their own government ranks in order to address the environmental challenges that come with economic development, from water to waste management, was critical. To aid in that effort, Admiral Prueher and I organized the first four-star-level Defense Environmental Conference, shortly after he took command of the enormous US Pacific Fleet.

This gathering, held in Hawaii in September 1996, included environmental and military officials from thirty-two Pacific Islands nations. By bringing officials from the US Environmental Protection Agency and the US Department of Energy to a defense environmental forum, we were modeling the behavior that we wanted to encourage among other nations. Environmental, energy, and military professionals were all working together, along with nongovernmental stakeholders from outside organizations. This kind of cooperation would become essential as Pacific Islands nations faced increasingly grave threats from climate change.

## Sounding the Alarm on Pacific Climate Threats

Our work together on environmental security while Admiral Prueher served as commander of the US Pacific Command, and his subsequent service as US ambassador to China, made him painfully aware of climate risks in the Pacific. Even back in 2007, Admiral Prueher could see that the effects of climate change—from flooding and extreme weather to glacial melting on the Tibetan Plateau—would lead to instability across the region. Referencing low-lying regions where arable land would be lost, he noted, "You see mass destruction in countries where the government is not robust. When people can't cope, governing structures break down," or people are willing to accept governments exercising "very firm control."[9]

In many ways, Admiral Prueher's foresight into the Pacific's future was prescient. He knew it was not possible to reduce global carbon emissions without China; we would have to engage Chinese leaders, even when there were fundamental disagreements. "Not talking to the Chinese is not an option," he said in 2007.[10] Admiral Prueher reiterated that message in congressional testimony that same year, arguing that the difficulty of working with China and India could not be an excuse for inaction.[11] Like it or not, these countries would be central to solving the world's biggest challenges. For Admiral Prueher, a proactive approach was critical not only to engaging allies and adversaries but to all aspects of climate security. Through his lens as commander of the world's largest military forces, trained and ready for a future fight in the Pacific, he observed:

> For military leaders, the first responsibility is to fight the right war, at the right time, at the right place. The highest and best form of victory for one's nation involves meeting the objectives without actually having to resort to conflict. . . . That seems to be a reasonable way to think about climate and security. There are a great many risks associated with climate change, and the costs are uncertain. But if we start planning and working now, we may be able to meet our security objectives, and mitigate some of those battles. The potential and adverse effects of climate change could make current changes seem small.[12]

Admiral Prueher was not alone in the navy in his concern about climate security. In 2013, one of his successors, the aforementioned commander of the US Pacific Command Admiral Samuel Locklear, advised Congress that climate change was the biggest long-term security threat to the Asia-Pacific region.[13] His direct experience in responding to the severe flooding that gripped Jakarta, Indonesia, in January of that year had clearly focused his attention on extreme weather and climate conditions in the region he commanded. Locklear could foresee that US forces would increasingly be called on to provide humanitarian assistance and disaster relief across the Pacific region as warmer waters and stronger storms destroyed communities and coastlines. For Locklear, climate change added "strategic complexity" to his mission to stabilize the region and could even threaten entire nations.[14]

Here was one of the nation's premier warfighters, commanding America's enormous forces stationed in Hawaii—designed to project power across the Pacific, deter China and North Korea, and keep the peace across a vast seascape that stretches across twelve time zones—telling Congress that climate was a top-tier threat to stability of the Indo-Pacific region.

That statement did not sit well with some of the most conservative members of the US Senate Committee on Armed Services. Senator James Inhofe of Oklahoma, then one of Congress's staunchest climate deniers, questioned the idea that climate change could be a significant security threat, let alone the biggest threat to the region.[15]

Admiral Locklear had not become one of the nation's top military commanders by avoiding difficult conversations. He stood his verbal ground, including under oath in the Senate. Acknowledging the threat posed by China and North Korea in the region, Admiral Locklear explained how geography and demographics make the Pacific region so highly vulnerable, even with its many highly developed economies. He noted that about 80 percent of the region's several billion people lived within two hundred miles of the coast and that the trend was increasing as people moved toward the economic centers. Those coastal areas are becoming increasingly vulnerable to rising seas and extreme weather events, such as cyclones and typhoons, which can in turn increase unrest and instability. "Natural disasters have second and third-order impacts on security," he explained. "These impacts include adversaries exploiting the stability created by a natural disaster, and internal unrest caused by food shortages and other domestic concerns."[16]

## Soft Power in Disaster Alley

What do these disasters look like? As Admiral Locklear noted, hundreds of thousands of people die every year and many more are displaced because of natural disasters, some of which are climate fueled. From 1970 to 2018, approximately 1.1 million people were killed in storms, floods, and other disasters (excluding earthquakes and tsunamis) in the Indo-Pacific region, and people there are five times more likely to suffer natural disasters than elsewhere.[17]

Coastal megacities such as Mumbai, Bangkok, Jakarta, and Ho Chi Minh City, where tens of millions of residents live less than one meter (about one yard) above sea level, are highly vulnerable to rising seas, storm surges, and saltwater intrusion into freshwater aquifers less than a meter above sea level.[18,19] China's Pearl River Delta, highly populated and home to the country's major economic engine, for instance, is at severe risk of flooding.

China is also facing record heat waves, severe drought, extreme precipitation, and glaciers melting in its Qilian Mountains. These trends not only threaten homes and infrastructure; they threaten the nation's economy and hit its poorest citizens the hardest.[20]

Even without the effects of climate change, Asia is known as "disaster alley" because the continent is home to more than 90 percent of cyclone victims worldwide.[21] That is an amazingly high percentage of people at risk from cyclones. The Bay of Bengal, especially the mouth of the Ganges River, is one of the most cyclone-prone regions in the world.[22] Like the Mississippi Delta, these areas can provide good soil for farming; however, even small rises in sea level lead to extensive flooding, especially during monsoon season.

Admiral Locklear saw the devastation these storms can wreak when he provided relief after Typhoon Haiyan ripped through the Philippines in 2013, just a few months after he gave his congressional testimony about climate security. That mission became a potent symbol of the military operating on the front lines of climate catastrophe. Admiral Locklear directed more than thirteen hundred flights to provide relief to starving and scared citizens. He had not been trained to operate in a world altered by climate change, but he knew he could not do his job or serve his country without factoring the climate crisis threatening the vast Pacific region into his work.

As he thought about how to confront China's growing military capability, he also had to ensure that his many soldiers, sailors, airmen, and marines could respond to the next climate-fueled storm, flood, or catastrophe that could destabilize important allies in the region. While Admiral Locklear's forces were saving lives after Typhoon Haiyan, he kept an eye on another important country in the Indo-Pacific region that is regularly in the climate crosshairs.

## Bangladesh on Climate's Front Lines

Locklear was "astonished" when he went to Bangladesh. He had no idea that the nation's entire population—165 million souls, or roughly half the number of people in the United States—lives in an area smaller than Iowa. And during a good part of the year, much of it is underwater. The capital, Dhaka, is teeming with twenty-two million people and is widely considered the most crowded place in the world.

Locklear painted a haunting picture of life in Bangladesh during the rainy season: citizens shuttling from their homes to the major cities, camped out on the high ground by the side of the road. He recalled first learning about the famines and floods in Bangladesh while watching the nightly news as a child. Every few years, reports would come in about tens of thousands of people dying in these floods. "Just tragic. . . . And nothing we could do about it."

But as a naval commander, Locklear could do something. The US military began partnering with Bangladesh to try to help mitigate the consequences of a changing climate. These efforts were not without their critics. Locklear recalled, "I had smart people argue with me when I was a four-star that there was no evidence of rising sea levels."[23]

Yet the evidence of rising seas and the toll they've taken is incontrovertible. In Bangladesh alone, tens of thousands have already been displaced and forced to relocate. The situation is primed to get even worse. Just a one-meter sea level rise along the country's southern coast would mean a loss of around 17 percent of Bangladesh's territory and could result in over twenty million additional people displaced, which would become a major force of regional instability as people flee to higher ground.[24]

Retired major general Munir Muniruzzaman of Bangladesh has been sounding the alarm on climate security risks in his country for decades. The very origins of Bangladesh as a country can be traced to Cyclone Bhola, which hit Bangladesh (at that time East Pakistan) in 1970. This cyclone resulted in 300,000 to 500,000 deaths, making it the deadliest cyclone ever recorded.[25] Pakistan's inability or unwillingness to help its then-citizens led to a movement for Bangladesh's independence from its West Pakistan Punjabi-speaking majority. The young Muniruzzaman first served in the Pakistan Army to help his fellow citizens but quickly joined the Bangladesh military when it first formed. For General Muniruzzaman's son, Shafqat Munir, one of his most formative experiences was the next devastating cyclone that hit Bangladesh in 1991 when he was eight years old.

Munir shared with me vivid memories of the US Marines arriving in hovercrafts, in what was called Operation Sea Angel. Some of the troops were just returning from the Gulf War. It was one of the largest disaster relief operations carried out by the Marine Corps to date and made clear that the scale of climate disasters could overwhelm the Bangladeshi state. In Munir's words, "It doesn't remain a climate event. It gradually becomes a security event."[26]

General Muniruzzaman served for thirty-eight years in the Bangladesh Army, retiring in 2008 to head the Bangladesh Institute of Peace and Security Studies. His military career, like the careers of Admirals Gunn, Prueher, and Locklear, became increasingly tangled in the accelerating risks of climate change as more and more parts of Bangladesh flooded with more and more frequency.

The good news is that the nation has improved its early warning systems and has taken other steps to become more resilient. It has done so in a number of important ways, as documented by Joshua Busby in his book comparing various climate extremes and abilities of countries to recover.[27] Busby's argument is that in Bangladesh and elsewhere, improving state capacity, making institutions more inclusive, and providing international assistance can be determinative factors. While cyclones have occurred with increasing ferocity over the past fifty years, fewer lives were lost in the major cyclones of 2007 than in 1991.[28]

The bad news is that Bangladesh's southern coastline is still a disaster waiting to happen. Already, Bangladesh is seeing displaced people move from the low-lying southern coasts, which are most vulnerable, inland to the most crowded urban spaces farther north. Movement to major cities such as Dhaka has already overwhelmed surrounding towns and rural communities, so the influx of even more people is becoming unmanageable.[29] Yet during the monsoon season, because of riverbank erosion and tidal cyclones, an average of three thousand to four thousand people per day arrive in the Dhaka slums.[30]

This crush of people and rapid urbanization are contributing to crime—and for understandable reasons.[31] There is no social safety net in Bangladesh. When people lose their livelihoods, they don't have savings and the state is unable to take care of them. Many end up in criminal gangs. Even those who avoid gang life face hunger, homelessness, and a lack of education and must navigate all the related issues that go with displacement without state support.

Beyond its own internally displaced people, Bangladesh is also home to one million Rohingya refugees.[32] Fleeing from oppression in Myanmar, these refugees arrive in a deeply stressed Bangladeshi environment, making them highly vulnerable and exploitable, which also contributes to heightening political tensions in the country.

After a visit to Bangladesh in 2023, the United Nations special rapporteur highlighted a host of issues facing migrants in the area: climate and economic stress; the government's inability to alleviate that stress; pressures from migration to and within Bangladesh; and inadequate refugee camps that are vulnerable to fires, flooding, and other disasters worsened by climate change. With the nation poised to lose 20 percent of its landmass while gaining thirty million people, "the math does not add up," and some Bangladeshis are trying to leave the country. The problem is that they can move in only one direction—toward India—because India borders the nation on all three sides, except the south, which is the sea. But for decades, India has unilaterally fenced the border, and border guards have been shooting Bangladeshi citizens trying to cross over—leading to an average of one hundred people killed per year.[33]

The result is a pressure cooker set to explode. Imagine a climate trend that sends thousands of Bangladeshis fleeing to a border that Indian patrols refuse to let them cross. That's the brutal reality. General Muniruzzaman lamented that there is no mechanism to cope with this circumstance. The Indian state, like many other countries, he said, is in denial regarding climate-induced migrations or climate refugees. He said, "It will not be just a human security issue; it will be a catastrophe."[34]

**From Drought to Fires: A Warming Australia**

This frightening scenario is getting attention across the Pacific region. In April 2017, I was invited to Australia to speak at public meetings about climate change's effect on global conflict and instability. My visit occurred against a backdrop of Australia's growing climate extremes, from extended heat waves and catastrophic bushfires to record rains and flooding.

Since 2017, these events have continued to worsen in frequency and intensity. In 2019 and 2020, extreme drought in Australia gave rise to unprecedented bushfires, which burned twenty-four million hectares (ninety-three thousand square miles).[35] In 2022, record-breaking rain and floods in eastern Australia (in Queensland, the wettest February on record saw almost nine times the typical monthly rainfall) caused unmatched devastation, forcing hundreds of people to relocate.[36] Sea level rise has also been threatening both the Australian coastline and neighboring archipelagic nations.

Consequently, there has been a surge in humanitarian assistance and disaster relief operations, both domestically and regionally. This has put a strain on the Australian Defence Force, which means it may be less prepared to engage in combat operations or to respond to increasing regional instability—the latter also being amplified by climate effects.[37] These pressures are a particular concern to the United States because Australia is one of our essential partners, in part because it can help check China's growing influence. Indeed, the United States, Australia, and the United Kingdom signed a trilateral security agreement in 2023 that will assist the Aussies in acquiring nuclear-powered submarines.

Since my visit in 2017, the connection between three serious threats has become apparent: China seeks a more active, and sometimes aggressive, role in the Pacific region; climate risks are mounting; and Pacific Islands nations are in even greater economic and environmental need.

Australia now recognizes that its own security is directly connected to addressing these three challenges together. In November 2023, the Australian prime minister signed a treaty with his counterpart from the Pacific Islands nation of Tuvalu to provide security guarantees from

Chinese aggression, protect the nation from climate change, and provide a path to migration for citizens of Tuvalu in Australia. Prime Minister Kausea Natano said Tuvalu had requested the treaty to "safeguard and support each other as we face the existence of threat of climate change and geostrategic challenges."[38]

## Threat of Extinction

Tuvalu is not alone in facing these existential threats. Some Pacific Islands nations may literally become extinct in the coming decades as seas rise and they lose their sources of fresh water. Kiribati is one such nation, once considered a Pacific paradise. It consists of a group of thirty-three atolls located in the central Pacific Ocean between Hawaii and Australia. Picture stilt houses on the beach; twelve different words for coconuts, depending on their level of ripeness; fishermen in sarongs collecting shellfish at low tide.

Kiribati and Tuvalu could well be the first countries to be swallowed up by the sea as a result of climate change. Global warming is melting the polar ice caps, glaciers, and the ice sheets that cover Greenland, causing the oceans to rise. It is estimated that sea levels have risen by an average of 3.2 millimeters (about 1.3 inches) per year since 1993, according to the *Fifth Assessment Report of the Intergovernmental Panel on Climate Change*.[39]

Pacific Islands countries are uniquely vulnerable to this change in sea level, as well as to coastal erosion, El Niño events, extreme weather, and ocean acidification. They are now on the front line of what some call "nation extinction."[40] On the one hand, Pacific Islands communities have shown remarkable and long-standing resilience in the face of both physical and social risks. At the same time, many of these nations, in particular Kiribati, Tuvalu, and the Marshall Islands, are geographically isolated and remote; heavily dependent on foreign assistance; distant from global trade networks; and have limited access to fresh water, energy, and natural resources.

Figure 6-1. Stewart Causeway, Kiribati (photo credit: mtcurado)

As these nations become more vulnerable to climate threats, will it be the United States or China that provides the support they need for their future?[41] That is an open question today. China has been extending its reach in the region, prompting some nations to shift their diplomatic affiliation from Taiwan to Beijing. There was a time, when Lee Gunn was growing up in the Pacific Islands in the 1940s and 1950s, that the United States was the clear strategic partner. That is no longer true. China offers these nations ports and airfields and seemingly low-cost loans that sometimes put them further in debt. China wants to fish in their waters, send its tourists there, and gain more access across this region. Will the United States be there when these small countries, both geostrategically important and longtime allies, need help simply to survive?

Anote Tong served as president of Kiribati from 2003 to 2016, during which he became a global spokesperson not only for the climate risks to his nation but also for helping his people "migrate with dignity" to higher ground. I spoke with him in 2019 when he attended a gathering on improving early warning for Pacific Islands nations at the Woodrow Wilson International Center for Scholars in Washington, DC.

He told an audience eager to hear his stories from the front lines of climate change: "I don't want our people to be relocated as climate refugees. So people ask me, so how would you like your people to be relocated? As people. Who would migrate with dignity. What that means is that it would be a proactive response to what is happening. We don't want to remain victims."[42]

This concept of migration with dignity is both important and underappreciated. It refers to people's freedom to choose whether, where, when, and how they move, prior to being forced to do so by climate impacts. To achieve it, we need a broad understanding of all the interrelated forces that prompt people to migrate, from economics to policy. And we need approaches that preserve migrants' culture and identities. As climate change puts ever greater pressure on communities to relocate, former president Tong is trying to seize the initiative, even if not from a position of strength. The goal is agency—and the possibility that migration can become a positive alternative under hard circumstances.

Let us end with Anote Tong's words: "I have 20 grandchildren and this is why I am very serious about this issue, it's very personal. But I think it's also about the other grandchildren, not just in Kiribati. Because I am hearing now that what is happening with climate change is no longer simply an existential threat for countries on the front line, but for humanity as a whole."[43]

CHAPTER 7

# Imperiled Neighbors to the South

GROWING UP IN PUERTO RICO, Elmer Roman knew better than many how thoroughly the natural environment can shape you. From hurricanes to heat waves, water shortages to power outages, extreme weather made life on the island unpredictable, to say the least. Those vagaries of the climate never dampened Roman's affection for his home, but he did want to see the wider world and was drawn to the military—both as a ticket overseas and as a way to serve his country. What he could not have known, when he joined up in the 1990s, was how much of his service would be devoted to protecting his beloved island from an increasingly unnatural set of natural disasters. As it turned out, the skills he developed in his youth would be just the skills the US territory would need in order to respond to a changing climate, including one of the hemisphere's worst storms.

After earning his degree in mechanical engineering from the University of Puerto Rico, Roman joined the US Marine Corps, becoming an engineering duty officer reservist and, later, a navy dive and salvage officer. This is no easy job: the repair of damaged ships is entrusted to only the most skilled divers who also have the technical knowledge to perform complicated tasks underwater. If you are picturing a navy officer in scuba gear, strike that image from your mind. Instead, Roman was outfitted with just a hard helmet, his flotation device, and intricate gear so he could maneuver in damaged, tight, dark compartments under the sea. Roman lifted the British icebreaker HMS *Endurance* from the Falkland Islands in the early 1990s and two guided missile destroyers, the USS *John S. McCain* and USS *Fitzgerald*, in later years. These were dangerous missions, to say the least, with Roman and his team against the weather, the incredible forces of the ocean, and the technical difficulty of lifting a heavy ship out of the water and onto another vessel.[1]

These skills were further tested in 2000 when al-Qaeda suicide bombers attacked the guided missile destroyer USS *Cole*, killing seventeen American soldiers and wounding dozens more. It was Roman's task to dive to the depths of the ship, repair what he could, and tow away what he couldn't. After completing the mission, he became one of the navy's experts on anti-terrorism and force protection. He began to develop new strategies to protect ships from terrorist attacks, including the use of protective zones around ships in port, warning devices, and advanced sensor technology. Yet Roman's work in the navy was not only about fighting adversaries. It was also about building alliances. He eventually joined the Office of Naval Research, a heavyweight among federal research institutions that shares technical expertise with foreign nations. As a navy science diplomat, he was assigned to Santiago, Chile, where he led teams from several Latin American countries to strengthen international partnerships and improve the nations' collective naval capabilities.

Here, Roman realized that the wild weather he experienced as a child would become even wilder as climate change drastically altered the Latin American landscape. Alongside Chilean and Argentinean colleagues, Roman evaluated both Antarctic melting and mass deforestation across the Amazon. He championed the new mapping technology known as lidar (light detection and ranging), created in MIT Lincoln Laboratory, which helped Brazilian government leaders better understand the topography of their forests and improve their management. In Costa Rica, he met with researchers measuring changes in tree rings to study how rising temperatures were affecting growth.

On February 27, 2010, just as Roman was about to leave his position in Chile, a magnitude 8.8 earthquake and tsunami struck the nation, leaving over five hundred dead and scores missing. Buildings collapsed; people were trapped in cars and under rubble; blackouts blanketed cities in darkness. Roman led the US Navy's response, drawing on his salvage and rescue experience to aid local search efforts and keep Washington officials informed of the ongoing recovery. When he returned to the United States, humanitarian assistance and disaster management remained part of his portfolio as he worked with the US Department of Defense (DOD) to improve the military's technological prowess across myriad missions.

Then, in September 2017, a natural disaster struck close to home. Hurricane María ravaged Puerto Rico, plunging its 3.4 million residents into a state of desperation. Roughly 80 percent of the island's crops were destroyed, taking with them $780 million in lost revenue: a disastrous hit for a community already struggling with economic disparity and poverty.[2] With the destruction of the electrical grid, the entire island lost power—and it took nearly a whole year to restore power for all residents.[3]

The hurricane ultimately displaced 130,000 American citizens and killed 3,000, though this number is difficult to confirm.[4] With morgues overwhelmed and families desperate to see the bodies of their loved ones, the island endured a "crisis of cadavers" for two years, Roman recalled.[5]

When the hurricane struck, Roman was on the other side of the world, in Singapore, leading the heavy lift salvage operation of the USS *John S. McCain*, which had collided with a tanker in August 2017. While still working twelve-hour shifts ahead of Puerto Rico's time zone, he supported the recovery effort by helping connect critical DOD organizations with the government of Puerto Rico.

He was there to aid in the immediate recovery, but his heart sank as he realized the federal government's response would be woefully inadequate. The image of President Donald Trump throwing paper towels to anguished American citizens—Trump told residents they should be proud they had not endured a "real catastrophe like Katrina"—remained raw, particularly as thousands were left without access to food, water, or a sturdy roof over their heads for months after the hurricane's landfall.[6] For Roman, it was particularly painful to know that so much suffering could have been prevented if the US and Puerto Rican governments had adequately prepared for the storm, or at least gone to work in a coordinated manner in its immediate aftermath.

In 2019, two years after the hurricane, Roman was appointed secretary of public safety by Puerto Rico's governor. The recovery had stalled. Much of the island remained in ruins. Crises started to compound as crime ticked up with every new blackout and food shortage. As temperatures rose and the fall hurricane season brought more rain and damaging winds, life on the island became increasingly chaotic and dangerous.

But Roman had little time to address safety and security issues before he was unexpectedly thrust into new roles. Just a couple of months later, the governor who had appointed Roman was himself ousted as public resentment toward the government's feeble recovery efforts boiled over. Suddenly, Roman found himself serving as both secretary of state and lieutenant governor. He had never considered himself a politician or run for elected office, but he did know how to develop and execute successful missions, so that is what he set out to do. Most importantly, he knew that the Puerto Rican government would have to rebuild its citizens' trust, and that would require working closely with a wide range of experts and city leaders to make sure that the local population received the

Figure 7-1. Damage from Hurricane María in Rincón, Puerto Rico (photo credit: cestes001)

support it desperately needed. In the face of such recent and widespread failures, it was no easy task, and one he would never have imagined for himself, but it was a challenge for which he had inadvertently been training for his whole life.

## A Shared and Threatened Neighborhood

Puerto Rico, like Elmer Roman himself, sits at the intersection of many identities, and the island has a complex relationship with the mainland United States. After all, Puerto Rico is an unincorporated territory but not a state. Its residents are American citizens but have no voting power at the national level. The sense of being betwixt and between, related to but separate from the rest of the country, comes with the legal status of a territory, but similar dynamics play out in the broader region.

Americans often recognize Latin America and the Caribbean as part of our shared neighborhood, given their physical proximity and close economic and cultural ties to the United States. Yet that closeness brings friction; many in the region distrust the US government, largely because of our historical involvement in their affairs. At the same time, however, America is widely seen as an escape from local hardships. Migration and large-scale displacements are a constant issue, which will only intensify as more people flee environmental disasters, degraded landscapes, and government instability.

For our military, the region presents a distinct challenge precisely because of its tight connection to us. The US combatant commands that oversee areas of the world with more overt conflict and adversarial relationships tend to receive more resources and support. The US Southern Command (SOUTHCOM) often gets short shrift, according to Chuck Hagel, former secretary of defense. To him, this is a clear oversight: "We've got to be solid in our hemisphere. We've got to be smart about our own hemisphere because these are our friends, these are our real allies. These are all part of the Americas."[7]

If there is one thing I've learned from my travels across this region, it is that a stable and secure Western Hemisphere is critical to our homeland defense. When US regional partners experience hardship, there will be consequences on our side of the border. As the 2022 National Security Strategy (NSS) aptly put it, "No region impacts the United States more directly than the Western Hemisphere."[8]

Those impacts became sharply apparent in the early 2000s as forced human displacements were intensifying. During his tenure as SOUTHCOM commander from 2006 through 2009, Admiral James Stavridis saw a significant rise in migration northward. As he came to understand it, climate change was the catalyst.

"I saw again and again and again the massive destructive effects of great storms and hurricanes. All of it is tied so clearly to the heating of the oceans and to global warming," he said.[9] This endless cycle of extreme weather barreling through the region not only creates immediate hardship but also stunts the long-term growth and evolution of affected nations. "It destroys the ability to create a functional livelihood," Admiral Stavridis said. "And what does that lead to? It leads to instability, and, of course, to us in the United States. It leads to illegal migration to our borders."[10]

Indeed, Latin America and the Caribbean region are among the places in the world most vulnerable to global climate change. Sandwiched in between the warming Pacific and Atlantic Oceans, the region is battered by rising seas, increased droughts, stronger hurricanes, and more frequent fires and floods.[11] At the same time, ocean acidification damages marine life, which ultimately hurts fisheries and food security. And as the exquisite coral reefs that protect the coasts are eaten away, there is greater risk from storm surges.[12]

While the effects of climate change are most visible right on the coasts, rising temperatures and shifting rain patterns cause damage farther inland as well, not only from flooding but also from prolonged droughts. Most communities have little capacity to store fresh water, which means there is a shortage of drinking water, let alone irrigation for farmland. The region's thousand-mile-long dry corridor, which stretches across Costa Rica, Nicaragua, Honduras, El Salvador, and Guatemala, is particularly hard hit. Over the past three decades, drought has resulted in the loss of about $5 billion per year in agricultural yields across this parched region.[13]

Beyond fires, floods, and droughts, climate change is driving the spread of fungal diseases that do enormous damage to agricultural livelihoods and can lead to social and political instability. Take coffee. As a 2017 report from the Center for Climate and Security noted

> A severe outbreak of coffee leaf rust in 2012/2013 impacted half of the approximately 1 million acres of coffee cultivated in Central America. Occurring simultaneously with a global fall in coffee prices, it meant Guatemala, El Salvador and Honduras saw the total value

of their export coffee crops decline from about US$3.4 billion in 2011/12 to US$1.6 billion in 2013/14. The region saw more than half a million job losses and all three countries declared national states of emergency.[14]

When environmental degradation is this severe, it quickly seeps into every dimension of life, from economic hardship to threats from terrorist organizations. Without strong governance and physical infrastructure to confront these hazards, communities are left extremely vulnerable; they are in desperate need of profound action to mitigate climate change and improve resilience.

The region's leaders recognize this. At the 2021 United Nations Climate Change Conference (COP26) in Glasgow, Scotland, Barbados's prime minister, Mia Mottley, called out the projected increases in global temperatures as a death sentence for island and coastal communities. "For those who have eyes to see, for those who have ears to listen and for those who have a heart to feel, 1.5 [degrees Celsius of temperature rise] is what we need to survive," she pleaded.[15]

General Laura Richardson, who became SOUTHCOM commander in 2021, echoed those sentiments in a statement to the US Congress in 2023, observing, "This is one of the regions most impacted by climate change. Hurricanes, rising sea levels, flooding, and drought are causing grave harm to the region's health, food, water, energy, and socioeconomic development. Extreme weather events impact our partners' national security, displacing populations and increasing irregular migration already accelerated by TCOs [transnational criminal organizations] and insecurity."[16]

**From Control to Cooperation**

Over the past twenty-five years, the United States has developed a very different relationship with Latin America and the Caribbean region from the one that existed when I first joined DOD. When I arrived at the Pentagon in 1993, the Cold War was still a recent memory, and military leaders were accustomed to thinking of the region in terms of containing the spread of communism. DOD had spent decades confronting, and sometimes ousting, a variety of left-wing authoritarian regimes in Latin America that the United States feared would join with the Soviet Union to expand undemocratic and communist rule. Today, one could argue that the region is still a proxy for competition between the United States and a major foreign power—this time, China, whose economic interests in the area include mining, mineral extraction, ports, roads, and illegal fishing. But rather than toppling or installing regimes, the United States is now working to gain influence through cooperation with regional governments, in part by supporting their efforts to prepare for the ravages brought by climate change.

It is a mission no one could have imagined at the dawn of the twentieth century, when President Theodore Roosevelt hatched a plan for an infrastructure project that would come to symbolize America's presence in the region: the Panama Canal. France had begun digging the canal in the 1880s but had abandoned it as engineering problems and worker deaths piled up.[17] Still, Roosevelt recognized its massive potential: opening a direct passage between the Atlantic and Pacific Oceans would transform global trade, creating a shortcut that would allow ships to avoid the long journey around the tip of South America and slice 7,800 miles from an overseas trip from New York to San Francisco.[18] So in 1904, after helping the Panamanians defeat Colombian rebels, the United States acquired land from Panama for this passage. It would take another decade, however, and many lives lost to tropical diseases, before the first boat traversed the Panama Canal in 1914. There would be many more to come. The passage completely changed international shipping routes, and while countries around the world made use of the new shortcut, US control of its operations signaled both economic and geopolitical power.

In 1977, following almost seventy-five years of US control of the canal, President Jimmy Carter signed a treaty with Panama to transfer operation of this critical global passageway over the course of twenty years. When I visited Panama in 1997 in my role as deputy undersecretary of defense (environmental security), the days of America's commanding position atop the Panama Canal were numbered. The Panamanians were eager to impress upon my delegation that they were prepared not only to operate the canal but also to protect the vast natural treasures surrounding it.

The flyer announcing my visit proclaimed that the "US military's environmental stewardship in Panama is building a legacy of trust"; indeed, the trip was designed to highlight cooperation between the two nations on environmental conservation. One of my stops was at the Smithsonian Tropical Research Institute, where I observed US scientists working closely with their Panamanian counterparts to preserve the incredible biodiversity of the region—research that was funded, in part, by DOD (another example of defense–science diplomacy). In an elevator-like contraption, I scaled the jungle canopy and peered out over the tops of the largest trees I have ever seen, staring in awe at the unparalleled diversity of the tropical forest.[19]

While the connection between the forest's health and the canal's operation might not be immediately obvious, it quickly became clear to General Wesley Clark, the last SOUTHCOM commander stationed in Panama. (Once the United States transferred operations to Panama, SOUTHCOM headquarters were relocated to Florida.) From the palatial commander's residence, perched high above the canal, General Clark had a bird's-eye view of how environmental change was affecting the passage's operation. Controlled through a system of locks and fed by water from Gatun Lake, the canal depends on consistent rainfall. "If you didn't get 100 inches of rain a year, you couldn't operate those locks because the Atlantic and Pacific are at different elevations," General Clark said.[20]

That rain is becoming increasingly scarce as climate change dries out the inherently hot region—a situation that is exacerbated by farmers clear-cutting the dense jungle surrounding the canal, since with less vegetation comes less precipitation. It all adds up to environmental degradation that threatens the Panama Canal's ability to function, which in turn threatens an entire network of global trade routes. But to General Clark and subsequent SOUTHCOM commanders, the damage was an issue not just of economics but also of security. Although the canal is a powerful symbol, both the consequences of environmental change and America's changing relationship with its southern neighbors stretch far beyond Panama.

As extreme weather becomes ever more common in the region, the military's mission increasingly concentrates on combating the deadly consequences of climate change. When Admiral Stavridis assumed control of SOUTHCOM a decade after General Clark, he restructured the command to focus on preparing for humanitarian relief efforts and countering the lawlessness that tends to accompany extreme weather events. When a massive storm strikes, he explained, the government has fewer resources to devote to policing, and large gangs take advantage.

"It encourages the use of drugs," he continued, "as both an economic vehicle and a palliative to the awful conditions."[21] The situation is ripe for exploitation by drug cartels, which pressure vulnerable migrants to smuggle narcotics across the border, supplying the drugs that feed addiction in the United States.

Under Admiral Stavridis's control and distinctive vision, the navy created the US Fourth Fleet, which plies the waters of the Caribbean, South America, and the Eastern Pacific, both to counter narcotics and to provide disaster relief. The admiral considers it his most important contribution at SOUTHCOM.

These types of missions are not what typically comes to mind when we think of military engagement. Indeed, they are markedly different

Figure 7-2. The USNS *Mercy* on a humanitarian mission (photo credit: Christa Stoebner)

from the large land battles we are accustomed to seeing during war, either in Ukraine or earlier in Iraq and Afghanistan. But as the pace and intensity of storms have accelerated beyond the ability of many nations in the region to recover from each successive event, the response by US forces becomes ever more central to regional stability. In the past decade alone, the US military deployed twice as frequently to provide disaster relief following major storms. Across eighteen nations in its area of responsibility, SOUTHCOM has established sixty-six emergency and disaster relief warehouses, stocked with necessary supplies to provide to communities in need.[22]

To prepare for this type of engagement, and to share expertise with our southern neighbors, SOUTHCOM conducts humanitarian assistance and disaster relief exercises each year across the region. These trainings bring together forces from different nations to exchange knowledge, ensuring that our military holistically addresses the needs of affected communities. The military services practice search and rescue exercises and examine best practices for operating in changing climate conditions, from rising sea levels to storm surges. The collaborations represent a new phase of the US relationship with our southern neighbors and a new imperative to come together to confront a global threat.

## Building Resilience

When General Richardson visited Puerto Rico in 2019 as commander of US Army North, she was deeply troubled by what she saw. Pallets of unused supplies—bottled water, food, diapers, cots—were sitting in warehouses, some nearing their expiration dates. "Going there two years after Hurricane María had struck and seeing FEMA still operating in a huge operations center and seeing the blue tarps still on a majority of homes throughout Puerto Rico was pretty devastating," she said.[23]

Former US Army civilian affairs officer Max (Adams) Romero put it even more bluntly: "The lag time in our response and the lack of coordination should not be a reality in the United States." Access to clean water, food, consistent education, and reliable infrastructure services are human rights. "It's bad enough when our partners in the region don't have access to them and we can't respond, but when it happens to our citizens in places like Puerto Rico, it's especially embarrassing."[24]

After assuming control of SOUTHCOM in October 2021, General Richardson recognized that natural disaster and humanitarian relief was a top priority in her new command. She knew that leaders in the region understand their own needs better than anyone else, so she worked to develop strong partnerships and support local communities to prepare for disasters before they strike. For instance, the US Army Corps of Engineers is building bridges to withstand worsening floods and building roads to bring food from rural farms to urban markets, while US forces are helping nations adopt advanced technologies to predict storm movements, along with early warning systems.

These efforts, while certainly important for local residents, are also in America's national interest. After all, when our southern neighbors experience disaster and disorder, chaos quickly becomes regional. In the words of Guatemalan president Alejandro Giammattei, "If we don't want crowds of Central Americans going to other countries in search of a better life, we have to build walls of prosperity in Central America."[25]

Building physical barriers to stop the influx of migrants, by contrast, has proved to be a strategy that is thorny and inhumane. In 2021, after four years of a harsh crackdown on migration, a record number of 5,000 unaccompanied children were in the custody of US Customs and Border Protection, and another 10,500 were in the care of the US Department of Health and Human Services, as President Joe Biden took office.[26]

Addressing a historic jump in migration, President Giammattei has called for cooperation between the United States and its southern neighbors. "We need the United States to see us as their front yard, as their partner."[27]

One basic solution is to create, in partnership with our neighbors, accessible, safe, flexible, and sensibly regulated pathways for migration—a step that is critical for the success of our dynamic nation. Yet a problem this complex cannot be solved simply by reforming immigration policy. It requires a broader effort to build resilience in the region, which includes everything from constructing bridges to conserving forests to strengthening democracy throughout local and national governments. Such efforts will become the opportunity multiplier that General Richardson hopes to build in the region. As Elmer Roman learned in the wake of Hurricane María in Puerto Rico, it requires hard-earned trust. And as Secretary of Defense Lloyd Austin has said, "You can't surge trust at the eleventh hour; trust is something you have to work on every day."[28]

## Preserving Democracy

Building trust was at the core of Max Romero's mission as a soldier with the Special Operations component of the Ninety-Eighth Civil Affairs

Battalion. Working across Latin America from 2014 through 2019, he saw firsthand that if the United States did not build strategic relationships in the region, others would step in—often to nefarious ends.

"It's difficult for countries like Guyana or Colombia or Trinidad and Tobago to provide disaster relief programs, infrastructure renewal services, and medical attention needed by these communities. So that ultimately creates resource gaps in the population, which can be exploited by nefarious actors, both state and nonstate in nature," he said.[29]

Romero's battalion gathered information on the specific challenges threatening these various populations and connected vulnerable communities to necessary resources. In Guyana, he and his team helped coordinate logistics after major storms washed away vital waterways in the Amazon Basin. The vast majority of Guyana's shipping comes through the Amazon River's inlets and riverways. As climate change exacerbates coastal erosion, these waterways can wash away overnight, making it impossible to ship goods along standard routes. These changes can be very difficult and expensive for coastal communities, Romero said, and can open up opportunities for illicit goods traffickers to take control of the passageways. By offering aid, Romero and his team helped stop traffickers from exploiting weaknesses as a means to assume control.

Over the course of his army career, Romero saw a shift in the US military's focus from fighting terrorism to competing with rival global powers, most notably China. This emphasis is particularly important across Latin America and the Caribbean, where China increasingly works to take advantage of the region's rich resources. From fish and forests to critical minerals and oil, China is trying to gobble up as many materials as possible, often illegally, through a multitude of unlawful mining, logging, and fishing operations.

These biodiversity crimes have devastating consequences for the region, General Laura Richardson said. Each year, between 350 and 600 ships subsidized by China fish off the coast of South America. With little to no respect for environmental regulations, the Chinese vessels are severely depleting fish populations, harming coastal economies, and destabilizing communities. "Not only do you have a drought corridor that is impacting countries' ability to feed their people but also, on top of this, the PRC [People's Republic of China] is raiding the fish to the tune of $3 billion in lost profit to the coastal countries," General Richardson said. "It just adds on to the recipe of insecurity, instability, and increased migration."[30]

Beyond stripping fish from the oceans and trees from the forest, these activities can pollute the environment. For instance, the chemicals used in illegal mining poison the water supply, making it doubly hard for local governments to provide for their populations.

Even sanctioned activities can have profound effects on the region. While supporting under-the-table operations, Beijing is also using legal, though hardly aboveboard, economic and political maneuvers to establish its power. Amid the 2008 financial crisis, China crafted deals to import oil, iron, soybeans, and copper from Latin America, buffering the area from the worst of the global downturn. As oil prices dropped in 2011, China once again rushed to the aid of struggling Latin American countries, presenting additional trade deals that proved irresistible at a time of financial instability.

In short, by flexing its economic muscle, China has established itself as a central ally to Latin American and Caribbean nations and is working to overshadow the US presence in the region. The long-term results not only threaten America's position in the world but also directly harm the countries that enter into these dubious arrangements. One key example is in Ecuador, where a Chinese corporation constructed the Coca Codo Sinclair hydroelectric facility, the largest energy project in the nation, at a cost of $2.6 billion.[31] Over the course of its construction, China provided $19 billion in loans to Ecuador to support the dam and to develop bridges, highways, irrigation systems, schools, health clinics, and a half dozen other dams across the country.[32] In return, China receives 80 percent of Ecuador's most critical export: oil. Extracting enough oil to repay Chinese loans has forced Ecuador's government to drill deeper into the Amazon, the world's richest tropical rainforest. "China took advantage of Ecuador," said Ecuador's energy minister, Carlos Pérez. "The strategy of China is clear. They take economic control of countries."[33]

As those countries become more vulnerable because of climate change, China will have ever more opportunities to assert control—unless the United States and other nations work to check its power plays and counter its autocratic agenda.

"Quite a bit of the challenges I encountered during my time in service were exacerbated by shifts in climate patterns, which, ultimately, just really magnifies the effects on vulnerable communities," Romero said.[34]

## Environmental Stewardship

In the waning days of US control of the Panama Canal, General Clark saw the indelible consequences of environmental change play out in nations throughout Latin America and the Caribbean. Traveling the region as SOUTHCOM commander, he witnessed the impacts of the ozone hole as tomatoes blistered on vines from overexposure and residents were unable to step outdoors in the middle of the day. Flying over snowless peaks in the Andes in the dead of winter, he had the stark realization that climate change will alter every corner of the globe, with severe implications for all of civilization. He recalled a true shift in thinking as he and other military leaders came to believe that as a function of the government, the armed forces have a responsibility for environmental stewardship. "You can't slash and burn, you can't trample the undergrowth, you can't recklessly clear land or pollute streams without having larger consequences," he said.[35]

That growing commitment to conserve natural resources has often been at odds with the hemisphere's largest nation, Brazil, home to much of the Amazon—a biodiversity treasure and critical tropical forest carbon sequestration.

While in Manaus, the capital of the state of Amazonas, General Clark met his four-star counterpart in the Brazilian military. The officer explained that he was working assiduously on his tactical military expertise to hold off aggression in the Amazon. When General Clark asked him who would attack the rainforest, he replied that the United States would. "You're environmentalists," he told General Clark. "We know this is ours and you're going to try to keep us from developing it."[36]

When General Clark arrived in the nation's capital city, Brasília, military leaders explained to him their plan for sustainable development of the Amazon, sparking General Clark to consider what, precisely, "sustainable development" referred to. "Is it sustaining the Amazon or sustaining the development?" he thought. After all, capitalizing on natural resources is not the same thing as preserving them. "It was a pretty shocking briefing," he recalled.[37]

General Clark knew then that militaries in the region would benefit from closer relationships with environmental agencies. Working with—and sometimes tangling with—the US Environmental Protection Agency (EPA) had certainly shaped thinking within the US armed forces. (General Clark recalled that early in his career, while he was stationed at Fort Carson in Colorado, an EPA inspector observed his motor pools for oil spills. The problem, he said, was that in that era, driving a tank meant spilling oil.)

So General Clark and I, along with the undersecretary of state for global affairs, former senator Tim Wirth, held a conference to foster new partnerships for environmental protection. Hosted in 1997, the conference drew together security and environmental leaders from across the Americas to share best practices in natural resource protection by militaries of the region. It marked a major new approach to conservation, promoting a sense of environmental stewardship among regional armed forces.

Since that first gathering, much progress has been made, with SOUTHCOM leading many climate education and training courses. In 2023 alone, I spoke to three different groups of military officers and foreign policy professionals about Latin American climate security, during talks hosted by the US National Defense University's William J. Perry Center for Hemispheric Defense Studies. Today's military leaders in the region are hungry for tools and techniques to climate-proof their societies.[38]

Meanwhile, leaders such as General Richardson and her SOUTHCOM team are performing the vital work of deterring threats from China and other malign actors in the region and helping the Western Hemisphere become more resilient to climate change. That is the work of converting the threat multiplier of climate change to an opportunity multiplier for action. Storms, floods, droughts, and other extreme weather events are, after all, both disasters that require immediate action and long-term threats to stability. For the United States and its southern neighbors alike, climate resilience is an all-hands-on-deck mission.

CHAPTER 8

# Climate Readiness on the Home Base

WHEN SUPERSTORM SANDY MADE LANDFALL in Brigantine, New Jersey, in the fall of 2012, Rear Admiral Ann Phillips was serving in the final command of her thirty-one-year navy career. As head of Expeditionary Strike Group 2, she was responsible for the US Navy's entire East Coast amphibious expeditionary fleet—fourteen ships and ten subordinate commands—which, among many other duties, could be called to support the US Coast Guard during recovery missions. But this was no typical recovery. As Sandy tore into the coastline, it flooded homes, hospitals, roads, and bridges across the Eastern Seaboard. US Coast Guard Sector New York on Staten Island and Coast Guard stations across the New York area were severely damaged by loss of power and flooding—in particular the entry to New York Harbor, which was managed by Coast Guard Station Sandy Hook. In an unprecedented move, the Coast Guard closed the entire Port of New York and New Jersey on October 28 and did not reopen it until November 4. At the moment the Coast Guard was needed most to conduct search and rescue operations, it was scrambling just to keep the lights on. Admiral Phillips recalled, "Coast Guard Station Sandy Hook was devastated. They were not able to operate. They had no power; their piers were destroyed."[1]

As directed by the US Fleet Forces Command and the Naval Expeditionary Combat Command, overseeing the response, Admiral Phillips's navy sailors stepped up to fill the void, loading amphibious ships at their base in Virginia Beach with generators and other equipment to deliver to the destroyed harbor in New York. Yet while the sailors' own naval base was still functional, it had been effectively sliced in two when Little Creek, for which it is named, overflowed its banks. Supplies bound for New York had to take a circuitous route just to get to the ships, traveling along the base's fence line, leaving and then reentering, before they could be carried to a landing craft, which would finally take them to the "amphibs."

Riding the last LCAC (Landing Craft Air Cushion) out to the USS *San Antonio* before it traveled through the night—on Halloween, no less—Admiral Phillips brought bags of candy so the crew could enjoy a short moment of levity. She returned to shore as the ship left port, and while the storm raged on, she and her husband maneuvered up and down the base's flooding roads in their personal vehicles, checking to ensure that all the ships belonging to Expeditionary Strike Group 2 were moored properly and prepared for riding well during the storm. Ships do not typically remain in port during a hurricane, but Sandy's force strengthened faster than expected, and as the storm moved slightly farther from the coast, there was no time to move the ships and be certain they could get ahead of the storm.

Meanwhile, the ships that made it to New York effectively became a floating temporary base where, for several days, the Coast Guard conducted all its operations, from managing vessel traffic to overseeing the port's recovery. "The Coast Guard moved as much of their command-and-control infrastructure as possible out to the amphibious ship

Figure 8-1. Marines and sailors off the coast of New York City in the aftermath of Superstorm Sandy, 2012 (photo credit: Corporal Michael S. Lockett)

USS *Wasp*, and they controlled the harbor from there until they could get their forces up and running," Admiral Phillips said.[2]

Even after the initial crisis had passed, the massive recovery effort dominated Phillips's fleet for weeks as her vessels and service members worked to simultaneously support New York and get their own base back on track. "We were not able to execute our duties in the way in which we normally would have because of the flooding," she said.[3] And that was just the immediate aftermath: over the longer term, there were effects on the natural environment; on the military's training regimes; on the surrounding communities; on the ability to acquire essential resources for the base, including energy, water, and food; and on the installation's infrastructure. "It's a series of cascading casualties with increasing significance over time, and *we are behind the rate of change*," Admiral Phillips stressed to me.[4]

The monstrous storm not only altered the course of Admiral Phillips's career; it was the first of several shock waves that would compel the US Department of Defense (DOD) to confront the impact of climate chaos on its own forces. For many, Superstorm Sandy was an inflection point. Its widespread destruction exposed the military to the same vulnerabilities we all face as seas and temperatures rise, storms and floods intensify, and droughts and fires become the national norm.

This onslaught of extreme weather has shifted military thinking about environmental considerations, even on its own bases. Over the past four decades, the military has moved from reluctant compliance with what were often viewed as burdensome environmental laws to an awareness that environmental protection improves military operations and contributes to a healthy quality of life for military service members and their families. I saw the beginning of this transition firsthand during my DOD tenure in the 1990s as the department began to adjust everything from the way troops trained (learning to take a more proactive approach to preserving natural habitats and ecosystems) to the use of toxic chemicals on bases. Today, the challenge is not only to preserve the natural environment within the base but also to ensure that its operations are contributing as little as possible to climate change—while preparing for the next Superstorm Sandy or whatever extreme event the warming world throws our way.

## The Environmental Back Forty

My experience while visiting military bases in the 1990s offered a peek into a unique world. I would arrive at the main gate, where military police would greet me, having already cleared me through complex security procedures. As the car drove past security onto the post, a signboard, used to announce movie and bingo nights, would display "Welcome Ms. Goodman," a friendly gesture to visiting DVs, or distinguished visitors.

I typically traveled with a military aide who had preplanned much of the visit with the installation's staff. I was blessed to work with many talented military officers during my eight years of service in DOD. Captain Joe Sikes and Major Tom Morehouse taught me military protocol in my earliest days, reminding me not to say "Sir" to someone more junior in rank because that would undermine my authority. Colonel Kurt Kratz and Major Tracey Walker accompanied me on many visits both in the United States and overseas, while navy captains Rick Rice and April Heinze kept the office, and me, on schedule.

Once I entered a base, I would be greeted by the commander, who would then show our team around the installation and update us on important developments in energy use and environmental safety. When most Pentagon officials visit military bases, they're proudly shown the site's newest weapons, from the US Air Force's latest fighter jet to the US Army's Apache helicopter outfitted with night vision goggles, new technology in the 1990s. None of that for me. (I did, however, spend my first day at DOD on a nuclear submarine, thanks to the always attentive head of the Naval Nuclear Propulsion Program, Admiral Bruce DeMars.) Instead, my tours, with the exception of one passenger ride in an F-16 fighter jet, consisted of what we called the "Back 40."

On one visit, I was treated to an up-close look at a waste dump that had become a Superfund site in the 1980s and, at the time of my visit, was being cleaned up through innovative treatment technology. On another, I was shown a wastewater treatment plant that had been upgraded to meet federal and state clean water standards and a range where unexploded ordnance from live-fire training was being removed.

I also observed how the military was conserving the natural and cultural resources of its bases. At Marine Corps Base Camp Pendleton in Southern California, the last stretch of undeveloped coastal land between Los Angeles and San Diego, I saw scores of protected and endangered species making their homes at the base. When I first met General Anthony Zinni in 1996, he was commander of the legendary First Marine

Figure 8-2. F-16 ride at Ramstein Air Base in Rhineland-Palatinate, Germany, October 27, 1993 (photo credit: author's personal collection)

Expeditionary Force at Camp Pendleton. The oldest, largest, and most decorated marine expeditionary force, One MEF, as it is known by marines, is a force of twenty-two thousand men and women organized to conduct complex and critical combat operations. While the image of this fighting force might call to mind brutal training drills, General Zinni observed that the fragile fauna and flora of Camp Pendleton required the marines to carefully preserve the natural resources surrounding them. "We had nineteen endangered species in Camp Pendleton, and it was restrictive in some ways as to what we could do," he told me.[5] In crafting their training regimens, the marines learned to accommodate their environment while ensuring the success of their various missions.

In South Dakota, I visited the reservation of the Oglala Sioux Tribe. The military had used the reservation as a training range for decades, contaminating local water supplies and making some sections off-limits to tribal members. I was there with a commitment to improve range cleanup and was greeted with a tribal ceremony and a traditional blanket, a sign of welcome and trust. At Wright-Patterson Air Force Base in Ohio, I toured the Wright Brothers' flying field and admired the unique Art Deco architecture of the buildings, which had been conserved with the help of the National Trust for Historic Preservation.

Whether my visit featured a historic building, a natural wonder, or a dump site, the most telling part of the trip often came in the evenings, when I had a chance to meet the spouses and children who made their homes on the base. When I visited Camp Pendleton, for example, General Zinni and his wife organized a dinner social with military families. I marveled at how effortlessly the spouses pulled together a delicious buffet, only later learning that it was the result of intense training—years of planning special events, moves, school changes, and doctors' visits, with as much dedication to detail as a war plan.

Military spouses, many of whom are women, are often as highly educated as their serving partners, but until recently their efforts have gone underappreciated. During the past decade, organizations such as Blue Star Families have advocated for professional development and employment opportunities for spouses, as well as the needs of military families. Those needs may change along with the threats from a changing climate—one more challenge posed by our warming world.

Traditionally, when service members deploy to conflict zones, their families remain home and out of harm's way. In the face of extreme weather, however, military families are no longer protected from the far-reaching impacts. Navigating the damage of Superstorm Sandy, Admiral Phillips witnessed the profound responsibilities placed upon military spouses. As sailors and marines were out at sea, the myriad bases stretching across Norfolk, Virginia, faced severe damage, leaving spouses and families to pick up the pieces, from flooding to school closures and lack of childcare. "You get the ships out of the way, but the surrounding infrastructure is all still there, and families are there," Admiral Phillips said. "All of your spare parts, your long-term fueling capacity, many of the things that you need to remain operational are left behind."[6]

For service members to continue to protect the nation, we need to protect their families, and the bases that make their missions possible, from the impacts of climate change.

## Growing Environmental Awareness on Base

Keeping everyone safe on military installations is one of the many responsibilities of the base commander. In many ways, the position is like being mayor of a small city, overseeing everything from family housing to natural habitats, barracks to bombing ranges, firehouses to commissaries, and roads to entrance gates. And while base commanders are typically colonels or navy captains—seasoned warfighters with at least fifteen to twenty years of service—most have not run municipalities before; that requires an entirely separate training regimen.

To keep them up to speed, DOD conducted an annual conference for installation commanders during the 1990s. One of the classes with almost perfect attendance was on environmental law and requirements, which I had the pleasure of teaching. The large audience focused intently as I spoke, both because they wanted to do the right thing for the environment and because they wanted to stay out of legal trouble.

As environmental regulations, at both state and federal levels, became stricter, commanders increasingly worried about criminal liability. Indeed, the criminal conviction of several officials at Aberdeen Proving Ground for dumping toxins in violation of the Clean Water Act in the late 1980s cast a long shadow.[7] "Knowing that you were going to be put on report for a hazardous material spill got a lot of people's attention," Admiral Phillips said, in the context of naval operations.[8]

And while dumping chemicals is an extreme example, it is easy to cause environmental harm without realizing it—and to run afoul of regulations. Take, for instance, Vice Admiral Dennis McGinn, who in the early 1990s commanded the aircraft carrier USS *Ranger*, docked in San Diego, and was issued a violation by the California Air Resources Board because of paint. Five cans of paint used to control corrosion had been left uncovered on the pier, which meant they were venting volatile organic compounds. As the commanding officer, Vice Admiral McGinn was subject to a fine and a black mark on his record.

It was a major lesson. "It basically reinforces the point that business as usual was not sustainable," McGinn said. "It was not good for the environment, the air quality, the ocean quality: we had to do things differently."[9] Later in his career, McGinn developed a training program to help sailors comply with environmental regulations and learn more about the importance of these laws.

Yes, the original impetus for programs such as these may have been to stay out of hot water. But by and large, the uniformed military, from installation commanders and ship captains to cadets, grew to realize that environmental protection also protected America's soldiers and missions. Today, their challenge is far more nuanced than simply following the letter of the law regarding environmental protection. It is a matter of navigating an ever-changing and increasingly complex natural landscape and building resilience. Rising sea levels at coastal naval stations, collapsing permafrost at ballistic missile sites in Alaska, raging wildfires encroaching on military fly zones: for our armed forces, climate threats are striking close to home.

Admiral Phillips was certainly not the only military leader whose worldview was shaken by the might of Superstorm Sandy. According to former secretary of defense Leon Panetta, the hurricane "brought climate change directly to the DOD."[10] Sandy paralyzed the entire East Coast as tunnels and roads flooded, power outages cloaked cities in darkness, and people were forced to flee the piles of rubble that were once their homes. The damage was so severe that only one entity could provide the massive relief required: the US Department of Defense. Panetta received calls from New York governor Andrew Cuomo and New Jersey governor Chris Christie requesting military assistance to restabilize their states in the aftermath of the storm's destruction.

"We were using our war capability in a war against nature herself," Panetta told me, pausing in a moment of reflection as we spoke, images of submerged cars, broken buildings, and flooded cities forever seared into his memory.[11] President Barack Obama rapidly mobilized the Hurricane Sandy Rebuilding Task Force, enlisting virtually every agency across the US government in an effort that emulated a "wartime council," Panetta recalled.[12] It was a pivotal moment for DOD; the impacts of climate change had become an enemy that required as much battle planning as any hostile actor. And Superstorm Sandy was just the beginning. The threats would only escalate as climate risks compounded.

As military personnel and equipment become increasingly devoted to responding to growing natural disasters, will they also have the bandwidth to prepare for other critical missions? Requests for military assistance with firefighting grew by a factor of twelve in just five years. As Deputy Secretary of Defense Kathleen Hicks observed, "The number of personnel days the National Guard spent on firefighting increased from 14,000 in fiscal year 2016 to 176,000 days in fiscal 2021.... And it is a major redirection of time, attention and resources."[13] According to the Center for Climate and Security's Military Responses to Climate Hazards (MiRCH) Tracker, in 2023 alone, US forces were called upon to squelch long-burning fires in Arizona, California, Canada, Colorado, Florida, Hawaii, Idaho, Montana, Nevada, New Mexico, North Carolina, Oregon, and Washington State.[14] National Guard and Coast Guard personnel and other first responders, as well as active duty military service members, have also rushed to support search and rescue efforts during various wildfires, including the devastating flames that engulfed the historic Hawaiian town of Lahaina.[15]

These wildfires are no longer confined to the summer months or to

Figure 8-3. Marine firefighter during the 2020 Creek Fire in Cascadel Woods, California (photo credit: Josh Edelson)

historical wildfire seasons. Daniel Hokanson, army general and chief of the National Guard Bureau, said that "the National Guard is now involved in fighting fires almost year-round."[16]

The fires not only take time and resources from the military forces but also affect their ability to train. In July 2018, wildfires damaged training grounds at Fort Cavazos (formerly known as Fort Hood) in Texas, forcing the cancellation of live-fire exercises, drills that are essential to ensure soldiers are battle-ready.[17] In other locations, such as at Fort Carson in Colorado, live-fire training itself has ignited wildfires.[18] And even in areas that have not been affected by burning flames and scorched earth, smoke has blanketed the land, darkening the skies and polluting the air over large swaths of the continent.

In addition to their environmental impacts, the enduring flames and their far-reaching smoke, together with prolonged heat and air pollution, create health risks for local communities and for troops as they train and deploy to fight wildfires. In the summer of 2023, heat domes covered most of the South and Southwest for weeks on end, smashing record temperatures in cities such as Phoenix, Arizona. At Luke Air Force Base, temperatures climbed above 110 degrees for a solid week. At times, the weather is so extreme that troops can't train at all, an occurrence known as "black flag" days.[19] Off the coast of Florida, ocean temperatures rose into the nineties, forcing navy ships to change their training schedules and routes.

These dog days of summer are increasingly punctuated by hurricanes that wreak havoc on critical installations around the nation, including Tyndall Air Force Base in Florida, Marine Corps Base Camp Lejeune in North Carolina, Offutt Air Force Base in Nebraska, Marine Corps Recruit Depot Parris Island in South Carolina, Naval Submarine Base Kings Bay in Georgia, and many more.

Together, these risks pose a triple threat for our military, affecting troops, training, and bases across the nation.

**Climate Readiness Is Mission Readiness**

Few recognize the threat climate change poses to the military better than Meredith Berger, assistant secretary of the navy (energy, installations, and environment) and the navy's chief sustainability officer. Alongside a team of navy leaders, Berger is laser-focused on preparing US sailors and marines to operate in all possible weather conditions—and to develop a "warfighting advantage" over our competitors as these conditions change. "Climate readiness is mission readiness," Berger stressed to me.[20]

To be truly ready, the military needs to build bases, develop equipment, and prepare forces that can withstand an evolving climate, "not just for the work to be done, but against the environment that we are building in," said Berger.[21] In 2021, DOD issued a Climate Adaptation Plan that requires "climate-informed decision-making" throughout the military, from knowing how high to build a floodwall to how much heat a soldier can endure to what type of energy should be used in future weapons systems.[22] The plan also directs each military base to prepare a site-specific installation resilience plan to identify local climate-related risks and mitigate them.

Of course, each base is different and faces distinct threats, in part because of its particular location. Not surprisingly, Berger is especially concerned about key coastal bases. In addition to the bases in Norfolk, Virginia, Naval Base San Diego in California and Marine Corps Recruit Depot Parris Island are among the early candidates for climate-resilient infrastructure.

As the primary training ground for new recruits, Parris Island in South Carolina proudly carries the motto "We Make Marines." It is a place rich with history and with a future imperiled by hurricanes, floods, coastal erosion, and extreme weather. The low-lying recruit depot experiences plenty of black flag days, with sweltering heat that interrupts the marines' training regimes. Without a plan to build resilience in the face of climate change, Parris Island could eventually become more a note in military history than a thriving base.

Fortunately, the base does have a resilience plan, which centers on the development of a microgrid to ensure the marines always have access to renewable, reliable, and redundant energy. Even if power goes down in surrounding areas, including as a result of extreme weather, the installation will stay up and running—and ready to act.

Investment in strengthening infrastructure and utilities that support critical missions has not always been a priority in military budgets, Berger said.[23] But this is starting to change. To ensure mission-readiness, today's military bases must be able to operate for fourteen days without energy from the electrical grid, and they will soon be required to operate for fourteen days with their own water supply.

Military leaders also realize that the source of the energy matters.

"Electrification is critically important as we think about emissions and the health and safety of workers," Berger told me.[24]

In addition, making military installations more energy efficient while able to perform the same tasks creates less strain on electrical grids, which also supply energy to surrounding communities; it is more cost-efficient, demanding fewer taxpayer dollars; it increases the energy security of the local communities; and it is less of a logistic burden. The Parris Island project will reduce the base's utility energy demand by 75 percent, limit its water consumption by 25 percent, and save $6 million annually.[25]

DOD, like many cities around the United States, is learning that improving energy resilience also reduces risks when the lights go out or the water is shut off. "If we can't turn the lights on or find the operational plans or know which direction to go, then we are at a disadvantage in the fight and we have not guaranteed that mission assurance that we are primarily responsible for," Berger said.[26]

The consequences can be grave, as John Conger saw while serving as assistant secretary of defense for energy, installations, and environment from 2011 to 2014. Conger witnessed two bases lose electrical power for more than three weeks: Naval Weapons Station Earle in New Jersey during Superstorm Sandy in 2012 and Misawa Air Base in northern Japan after the Fukushima nuclear disaster in 2011. These incidents and others like them prompted Conger and the DOD team to develop DOD's first Climate Adaptation Roadmap in 2014.[27]

Conger learned that while most military installations had electric generators, these weren't necessarily powerful enough to keep the bases up and running, and they weren't being tested for that purpose.[28] His early efforts laid the groundwork for the military's Black Start exercises, first conducted in 2019. These "pull-the-plug" tests determine whether the base can maintain power for fourteen days when nothing is coming from the electrical grid, in order to keep performing critical missions.

Not surprisingly, when you frame climate risks as a threat to the military's mission, ears begin to perk up. "I've found that, as you try to make a policy or a business case for investments in climate change or in energy resilience, you have to do so in a way that makes sense from the mission lens of the Department of Defense," army veteran and United States Military Academy West Point graduate Richard Kidd told me.[29]

As a logistics officer for the United Nations in the 1990s, Kidd came to understand the importance of fuel efficiency and access to natural resources in conflict management and peace building. Working in refugee camps across Bosnia, Rwanda, Tajikistan, and Afghanistan, he saw that poverty and lack of energy went hand in hand: without power, there could be no security, or medical care, or clean water. Similarly, while providing humanitarian relief across Cambodia, Laos, Vietnam, Angola, and Mozambique as part of the US Department of State's Bureau of Political-Military Affairs, he discovered that in war-torn countries, scarcity of natural resources often contributed to conflict and that environmental damage, whether from land mines or waste, stood in the way of post-conflict recovery. In 2007, he made the conscious decision to leave the State Department and join the US Department of Energy (DOE). "If I spent my life on post-conflict response, I was going to be on the wrong side of the equation," he said. "And the right side of the equation is helping to create models of sustainable development, of operations, of energy efficiency, of circularity, of rejuvenation and renewal."[30]

Driven by that goal, Kidd helped write the federal government's first executive order on sustainability, Executive Order No. 13514, issued in October 2009 by President Obama.[31] At that time, federal statutes (the Energy Policy Act of 2005 and the Energy Independence and Security Act of 2007), as well as the executive order, focused on improving the energy efficiency of buildings and vehicles.[32] To help implement this executive order, Kidd returned to his roots with one of the government's biggest energy users: the US Army. In response to the new federal directives, the army put forward some of the most advanced high-performance sustainable building standards in the federal government at that time, including energy efficiency standards and advanced materials. To understand how these standards were being received and implemented, Kidd visited several districts of the US Army Corps of Engineers. He found that the service members' opinions of the new rules varied widely, to say the least.

> I met the senior GS 15 (senior General Schedule), and he said to me, "There's nothing you can teach me. This standard means nothing. I've been designing barracks since the 1970s. I've got this." Could not have been more dismissive. And then the next day I ran into a young GS 13 with a master's degree in all sorts of LEED certifications, and she said to me, "I just can't believe we're building the same crappy barracks today that we built in the 1970s." You have those two perspectives concurrent inside the organization: one is actively resistant to change, and one can't get things to change fast enough.[33]

Which of these views will win out? Status quo or progress? Kidd thinks the answer depends on whether the military sees climate resilience as a part of, or a distraction from, its central mission: protecting lives and defending America. "The Department of Defense is very mission oriented, and it should be," he said. "It's very concerned about operational effectiveness and safeguarding the lives of service members, so anything that you add on that detracts from that mission is not welcome. But if you're able to bring solutions that help in that mission and reduce greenhouse gases and make the department more resilient, that's a good thing."[34]

Today, safeguarding military missions goes hand in hand with preparing for climate threats. Certainly, our forces must be capable of operating in any environment, but they also have to stay ahead of our competitors. Meredith Berger explained that adversaries are also developing energy efficient buildings and vehicles, which means less time and exposure while transporting fuel, as well as considering how extreme weather might be used to their advantage.³⁵

These moves are a serious concern for Berger and her colleagues, who recognize the necessity of employing the latest and best technologies to remain competitive. Today, warfare isn't only about who has the biggest tanks or the fastest fighter jets; it's also about who can improve their energy efficiency and thrive in a warming world. "When we put a sailor or marine on a task, they will go out and they will succeed," Berger said. "And they will, every single time, create an unfair advantage over the competitor that they're engaging with."³⁶

## Climate-Proofing Our Bases and Forces

Meredith Berger, alongside many others at DOD, is working to climate-proof our military against today's threats as well as the ones lurking down the road. And that means facing tough realities about the vulnerability of some military bases. As Richard Kidd put it, "We have to start to prepare now for hard decisions. We shouldn't invest money for fifty years in an installation that's not going to be there, or money in an installation that we're not going to be prepared to spend additional money to adapt. We need to really change our planning horizon on climate change adaptation and climate change investment."³⁷

No military base embodies this challenge better than the naval station in Norfolk, Virginia, an area with the highest rate of sea level rise on the East Coast. Although storms have always flooded Norfolk, they now occur with greater intensity and severity. Today, Norfolk is ranked tenth among the world's port cities most at risk from rising seas. A recent report by the North Atlantic Treaty Organization (NATO) observes that while ships can still leave the harbor and planes can still operate for now, there is no question that flooding already affects the US Navy's readiness.³⁸

Figure 8-4. The USS *Wisconsin*, docked in Norfolk, Virginia (photo credit: jfbenning)

Preparing her beloved Norfolk for this new reality became Admiral Phillips's calling after she retired from the navy. In 2018, she accepted an invitation from Virginia's governor, Ralph Northam, to serve as his special assistant for coastal adaptation and protection. In this role, she crafted a coastal resilience master plan for the Commonwealth of Virginia, bringing together eight planning districts and a multitude of stakeholders, including federal and local agencies as well as all four military departments and a NATO command—the NATO Allied Command Transformation.[39]

Admiral Phillips approached this mission in the same manner as any other. First, she assessed the greatest risks and vulnerabilities, including threats to underserved communities whose voices had not always been heard. The next step, determining building standards for everything from roads and rails to bridges, was no easy task because current regulations do not match today's climate realities. But the Commonwealth of

Virginia, under Governor Northam, issued executive orders to require the state to develop sea level rise projections and create standards to protect state property from flooding.[40] Rather than rely on outdated historical standards, Admiral Phillips set out to revise them, using predictive modeling to gauge future flooding, rainfall, and sea level rise. As she told me, simply patching up buildings from this year's storm without developing infrastructure that can withstand tomorrow's disasters means "we are always chasing our tail." Instead, we need to get ahead of the climate curve.[41]

To do so, the US Navy—with substantial support from Congress—is working to make the naval station, and Norfolk as a whole, more resilient. This includes adding berms and floodwalls to prevent erosion, retrofitting critical buildings, constructing new facilities that will weather future storms, and raising and widening a major pier to better accommodate submarines.

These changes are critical to ensure that Norfolk can continue to protect not only the United States but also the entire Euro-Atlantic area, according to NATO's report.[42] For the most powerful military alliance in the world to recognize climate change as a threat multiplier shows just how serious the situation is and just how far the conversation has come since the 1990s.

To safeguard the Norfolk naval base and the many other vulnerable military bases, Secretary Panetta advises becoming more imaginative about both potential climate threats and solutions. Rather than wishing for the past, we need to prepare now for a "future climate Pearl Harbor," he said.[43] Some national security professionals frame this approach as syncing up our "warning machine" of intelligence and prediction with our "action machine" of responding when disaster strikes.[44]

While military leaders are becoming more proactive when it comes to climate change, they are also following the adage "Never let a crisis go to waste." If Superstorm Sandy gave New York and New Jersey the motivation and the funding to make the region more resilient, Hurricane Michael did the same for Florida. In 2018, the Category 5 storm decimated Tyndall Air Force Base, just east of Panama City, destroying virtually all hangars for F-22 and F-35 fighters and damaging 484 buildings beyond repair.[45] Five years later, in 2023, the base was becoming a model of climate resilience as buildings, runways, hangars, childcare centers, and more were rebuilt to withstand higher winds, rising seas, and more intense rain and floods. "What Michael did for us is, it wiped the slate clean," said Don Arias, a spokesman at Tyndall's Natural Disaster Recovery Division. "It gave us the opportunity to reimagine."[46]

The new vision for the base is rooted in nature-based solutions. After Hurricane Michael flattened the dunes and seagrass meadows that abut Tyndall along the Gulf Coast, a major restoration project created a "living shoreline" of plants, rocks, and oyster reefs to prevent erosion and protect the peninsula. The effort was supported by The Nature Conservancy, which has been working with military bases since I signed the first memorandum of cooperation in the early 1990s. These natural features are complemented by high-tech innovation. For instance, the air force created a digital twin of the installation, enabling officials to simulate how new structures will fare as weather conditions become more extreme.

While Tyndall offers a hopeful story of renewal, each base requires a different strategy, and not all of the strategies are straightforward. Richard Kidd observed, "The effects of climate change are not going to be evenly distributed. And there may be some cases where DOD has to make some hard decisions" to protect the military mission.[47] Kidd laid out three approaches for bases facing increasing climate threats: defend in place; adapt the mission; or, at times, conduct a measured withdrawal.

Defending in place may seem like the status quo option, but it still requires adaptation. Take Joint Base Langley-Eustis in Hampton, Virginia, home of Air Combat Command and of F-22 Raptor fighters. Floodwaters frequently wash over runways at the base, preventing the pilots from taking off, but it took Hurricane Isabel to really get the attention of air force planners. In 2003, a 7.9-foot storm surge flooded the base for four days—wreaking havoc that could no longer be ignored. "The definition of insanity is to repeat the same thing and expect different results; we were not going to do that," said Robert Barrett, a base engineer. "After Hurricane Isabel, our Civil Engineers realized we needed to make some changes, and our journey of climate and installation resiliency began."[48]

From installing water pumps and cisterns on the runways to moving mechanical systems to higher ground to building organic seawalls, the air force has taken key steps to protect the base, which is critical to protecting the US capital. Given Langley's proximity to Washington, DC, the billions invested were clearly justified, Kidd said.[49]

At times, though, it makes more sense to change the mission than the base. Fort Stewart in Georgia, for instance, is currently the largest armored training facility on the East Coast and the primary base for the army's Third Infantry Division. But it can't remain that way forever. Sea level rise is turning the training grounds into swampland. While restoring mangroves and salt grasses can offer some protection for the base, eventually it will become harder for Fort Stewart to support heavy armored formations. At some point, the tanks and trucks will get stuck in the mud. "So maybe you keep Fort Stewart, but you take the armor out and you put light infantry in," Kidd suggested.[50]

For some bases, the outlook is bleaker, and changing up the mission may not be enough. At the two combatant command headquarters in Tampa, Florida, or at Naval Air Station Key West, climate change may prevail, and as Kidd noted, preserving these installations may not be fiscally responsible. Sometimes, closing the base and moving the mission is really the only option.

These are difficult conversations. No one at DOD enjoys the idea of abandoning military installations with deep histories and important records of service. And since no politician wants to lose jobs in their district, Congress has been unwilling to let DOD make these tough decisions and has not authorized another round of base closures since the early 2000s (see chapter 2).

America's leaders, however, have an obligation to this nation and its citizens to protect and preserve our national security. This mission will fail if the military cannot prepare its service members and facilities for future threats.

## Climate Leadership

A warming world is, in many ways, one of our fiercest enemies. "We've got to protect our country from this adversary called climate change," said Secretary Panetta.[51]

Beyond military installations, entire cities have been decimated by fires, hurricanes, floods, earthquakes, and storms. Recalling a California earthquake that occurred during his congressional tenure, Secretary Panetta was overcome with the memory of lives lost and upended. "Even an adversary could not produce so much damage," he said.[52]

This damage will only get worse.

As climate change increasingly seeps into all aspects of our lives and all communities across the globe, the US military holds a unique responsibility to assume a leadership role. Fortunately, with strategically important bases and training facilities around the world, DOD is well positioned to take up the mantle and develop climate-resilient infrastructure.

And it is doing so. Kate White, a DOD climate program director and veteran of the Army Corps of Engineers, has devoted much of the past decade to developing the DOD Climate Assessment Tool (DCAT), which draws from a multitude of global data and models to screen military installations for risks. "The department really has been thinking seriously the last few years of how to understand and plan for these future realities," she said.[53] Armed with that understanding, the department is using its buying power (evidenced by its $5 billion investment in a single base over seven years) to become an early adopter of advanced technologies and techniques.

By virtue of its sheer size, its large budget, and its mission-critical need to understand evolving climate threats, DOD holds profound power, said Admiral Phillips.[54] That means it can push the rest of the government to act now rather than wait for the next catastrophe. And it can begin to break through the partisan logjams that stand in the way of meaningful climate action.

After all, climate change does not discriminate between Democrats and Republicans. In fact, many of DOD's most vulnerable bases are in historically red states, given that so much of today's military is based in the South. Fortunately, the armed services committees of Congress have taken a largely bipartisan approach when it comes to providing guidance and funding to help our forces and bases become more climate-resilient. "There's got to be a sense that this is not about politics, it's not about culture wars. It really is about our security," Secretary Panetta said.[55]

While climate change has often unfortunately been politically divisive in the United States, there is broad support for protecting the troops that protect our nation. In bringing together diverse constituencies to face climate threats, the Department of Defense is again leading by example.

## A Bird's-Eye View of Military Readiness

The importance of bipartisanship became crystal clear to me when both houses of Congress swung from Democratic to Republican control in the 1994 midterm elections. This "Republican revolution" gave the party unified control of Congress for the first time since 1952. For President Bill Clinton, it was a major setback to his domestic legislative agenda, particularly his health-care plan. For me, less than two years into my tenure at DOD, it meant that the chair of the Senate armed services subcommittee overseeing DOD's environmental budget would be Senator James Inhofe of Oklahoma, a Republican, who had just narrowly clinched the election against Congressman David McCurdy, a Democrat.

Unlike Congressman McCurdy, who had personally invited me to help inaugurate a new environmental center at Fort Sill in his district, Senator Inhofe was distinctly hostile toward environmental matters. In his view, conservation equaled unwelcome restrictions on the oil and gas economy of Oklahoma. Environmentalists were either bureaucrats or activists, neither of which he held in high esteem. He did, however, have high regard for the military, especially for "his" military bases in Oklahoma, including the army's Fort Sill, Altus Air Force Base, and Tinker Air Force Base.

One of the early lessons I had learned while working in Congress and later at DOD is to appreciate the local lens through which any elected representative views the world. Even though Senator Inhofe would not naturally support environmental funding in the defense budget, I recognized that military readiness, particularly at Oklahoma bases, was a major priority for him. So I set to work on showing him how the military was promoting both operational readiness *and* environmental protection. There was no better way to observe this than with a bird's-eye view.

In June 1997, I flew with Senator Inhofe in a plane he personally piloted around his home state. Although the idea of winging through the skies in a tiny aircraft made me a bit nervous, I naturally accepted his offer to fly me to three Oklahoma military bases.

We visited Fort Sill, the primary artillery training base for the US Army and the US Marine Corps, to see how the servicemen rotated maneuver training fields just as a farmer would rotate crops. This practice helps create a realistic experience for the soldiers while preserving the natural resources of the tallgrass prairie. We visited Tinker Air Force Base, a major industrial depot and the largest single employer in Oklahoma, to observe how the air force conducted environmentally sound bead blasting to remove old paint from aircraft using nontoxic materials. At Altus Air Force Base, we learned how the installation was managing water resources in bone-dry land to support the needs of both the military and the local community.

Each visit reinforced that military readiness and environmental protection go hand in hand. Although our travels hardly converted Senator Inhofe into a tree hugger, I did gain his support for environmental programs in DOD. Flying throughout his beloved state, we both saw that conservation does not have to stand in the way of military training. It is not only possible but also imperative to achieve both. The experience demonstrated that, ultimately, politics is about people: if you can build personal relationships, you can often overcome the partisan deadlock that threatens both our planet and our military. As storms, fires, and other ravages of climate change become more deadly—on military bases and in civilian communities alike—we will need those relationships as never before.

CHAPTER 9

## *Less Fuel, More Fight*

AT THE HEIGHT OF THE US DEPLOYMENT to Afghanistan in 2008, US Army lieutenant Mike Henchen was responsible for all vehicles entering the country's largest American base, Bagram Airfield, in Parwan Province, about sixty kilometers (37 miles) north of Kabul. Housing forty thousand troops and civilians, the gigantic outpost was like a small city, complete with shops, Dairy Queen restaurants, and traffic jams. Yet this was no ordinary community. Henchen and his platoon had to verify that every car or truck passing through the base's gates was authorized to be there and then do a thorough search for explosives or weapons. After numerous suicide bombings, including one in 2007 that killed twenty-three people and injured twenty more while Vice President Dick Cheney was visiting Afghanistan, security on the base was on heightened alert twenty-four seven. Tensions ran high.

"Every week or two, there would be some little security scare, something that didn't look right," Henchen told me. If a single dog signaled that an entering truck smelled of explosives or the troops couldn't make out an object on the enormous X-ray machine that screened every vehicle, Henchen and his company had to lock down the entire entrance point until the explosive ordnance disposal (EOD) team gave the all clear sign.

"It was just this huge, huge to-do," he said with a sigh.[1]

More than three hundred vehicles entered the base daily. "All sorts of vehicles came in," Henchen said, "but the largest share of that was trucks carrying fuel. Constantly, fuel, fuel, fuel, fuel, fuel coming into the base."

After being inspected by Henchen and his team, some of this fuel was used for Bagram's all-diesel electrical grid, some was loaded into military vehicles stationed at the base, and some was transported to yet another military operation. From his position at the entrance, Henchen watched cargo helicopters carrying giant fuel bladders rise above Bagram as they set off to even more remote installations in the region. "It was just this enormous journey," he said.

Consider the path the oil took: After being extracted from the ground somewhere in the world and refined, it was loaded onto a tanker and transported to Karachi, Pakistan. From there, it was placed on a truck and driven across Pakistan, over the Khyber Pass, into Afghanistan. It then traveled rural roads through the eastern portion of the country before it finally arrived at Bagram, and there it was searched by Henchen and his platoon. The base might have been its final destination, or it might have taken one more journey, being flown off to the remote countryside.

This passage was long and winding, not only demanding a tremendous amount of manpower but also threatening the security of each individual along the fuel's intricate route. "My platoon was just one point along the chain," Henchen said. "We were probably the point along that whole chain that was most worried about getting blown up."

It was a constant fear, on every soldier's mind, during every inspection. This could be the truck that was carrying explosives. This could be the day that turned deadly. And so, tedious as the checks were, Henchen and his troops performed them with intense care and precision.

Living with the daily monotony, strained by anxiety, Henchen started to consider the immense potential of reducing our military's reliance on fuel.

"If we could actually change the way the military uses energy to become less reliant on this fuel, it would not only save money but would be a tremendous relief on the security risks that our soldiers were facing," he said.[2]

Senior defense officials were having the same thoughts, though changes had yet to trickle down to Henchen's platoon. Not only were America's sons and daughters in harm's way with each incoming fuel truck, but also the cost of getting energy to remote locations in Iraq and Afghanistan was becoming a substantial financial burden for the US Department of Defense (DOD). The price of delivering each gallon of gas to the front was a whopping $400 by the time the costs of transport and security were fully factored in.[3]

Disentangling our military's tether to fuel, however, is an enormously complex undertaking, requiring significant investments and technological advancements. But it's worth the effort for the benefits to the security of our troops, our nation, and our climate.

**A Dangerous Tether**

DOD is the nation's single largest energy user, accounting for about 1 percent of all US energy consumption and 76 percent of the government's fuel consumption.[4] A deep dependence on fossil fuels has long been woven into all military missions and conflicts. For decades, building the strongest, best-equipped military in the world seemed synonymous with using an enormous amount of oil and gas. Modern weaponry, global operations, keeping our forces safe and ahead of rival nations: it all takes energy—a tremendous quantity of it.

Figure 9-1. Refueling US Navy Seahawk helicopters (photo credit: sierrarat)

As the generals and admirals of the CNA Military Advisory Board noted in 2009, for our armed forces, energy is a double-edged sword.[5] Yes, it powers the tanks, trucks, aircraft, and weapons systems that safeguard the nation. But trucking fuel around the world, particularly in conflict zones, puts our troops in danger. And if a military base loses power or if fuel can't reach our troops, it's a serious security threat. From the logistic puzzle of moving fuel to the front for our long-standing engagements in the Middle East to our current reliance on the electrical grid to conduct unmanned missions, uninterrupted access to energy is mission-essential.

This historical reliance on fossil fuels, paired with the US government's strategic decision to ensure the free flow of oil to our nation and our allies, has shaped many military missions.

In 2003, Colonel Greg Douquet deployed with the Third Marine Aircraft Wing of First Marine Expeditionary Force (One MEF) as the lead air planner in the mission to invade Iraq and overthrow Saddam Hussein's regime. Colonel Douquet's job was to make sure that the initial "shock and awe" air strikes did not hit the region's oil fields.

"Our civilian masters had made crystal clear that intact oil production was essential to getting Iraq back on its economic feet after we knocked the stuffing out of them," Colonel Douquet told me. "We even met with civilian petroleum industry experts to incorporate provisions for shutting down flaming rigs, since Saddam had threatened to attack his own fields in order to deny them to us. That would have created an environmental disaster by dumping millions of barrels of crude into the Northern Arabian Gulf."[6]

While Colonel Douquet is grateful that DOD devoted extraordinary resources and considerable care to protecting the Iraqi environment he was sent to invade, he recognizes that this environment was a big part of the reason he was in Iraq in the first place. Although the US government declared that Hussein possessed weapons of mass destruction (WMD), the primary motivations for the invasion were Iraq's strategic location at the mouth of the Northern Arabian Gulf and the country's massive oil reservoirs.

"The protesters at home weren't wrong: we wage wars for oil," Colonel Douquet said. "Our global economy and way of life would collapse without it."

Indeed, the 2003 invasion of Iraq was the culmination of years of conflict over oil, starting with the first Gulf War in 1990. The war ended in détente enforced by US fighters based in Kuwait and Saudi Arabia, contributing, along with other factors, to the rise of Osama bin Laden, al-Qaeda, and the events of September 11, 2001.

Six years into the war in Iraq, the CNA Military Advisory Board observed that "our dependence on foreign oil reduces our international leverage, places our troops in dangerous global regions, funds nations and individuals who wish us harm, and weakens our economy." This dependence locked DOD into a vicious cycle of fighting conflicts to gain access to foreign oil—all while relying on that same oil to fuel the mission.[7]

As Colonel Douquet reflected upon his time in the military, he recognized that the last two decades of his service were, to a significant extent, dedicated to securing America's uninterrupted supply of Middle East oil.

"My focus was on the mission and protecting the marines and families entrusted to me," he said. "Yet when most of my deployments increasingly revolved around securing fossil fuels, I couldn't avoid the role the environment played."[8]

Today, we know that this dependence is not sustainable. Not only does our reliance on oil intertwine our nation with questionable regimes, it also inspires—and even funds—attacks on our own forces. As the CNA board wrote more than a decade ago, "Our nation's energy choices have saved lives; they have also cost lives."[9]

The consequences of those choices became painfully apparent to Ray Mabus when he served as US ambassador to Saudi Arabia, the world's largest oil producer, in the 1990s. A former governor of Mississippi, an oil-heavy state, Mabus witnessed firsthand oil's profound influence in shaping nations and their politics, militaries, and transnational alliances. US strategic interests in Saudi Arabia have long been tied to its role as the world's largest oil producer.

"One of the things that I took away was just how powerful, and also how inconsistent, oil was," he told me. "And that it could be used as a weapon for or against you."[10] He carried this perspective with him when he was nominated by President Barack Obama to serve as secretary of the US Navy.

In preparing for his confirmation in 2009, Mabus received consecutive weeks of military briefings in what he described as a "windowless conference room on the third deck of the Pentagon." These briefings were long and intense, Mabus recalled, with frequent references to our military's reliance on energy. "It kept popping up," he said, as both a vulnerability for the United States and an opportunity for weaponization by our adversaries. Our forces' constant need for fuel demanded complex logistics, as well as deals with foreign leaders, exposing our forces and our nation.

This energy burden could not be ignored, Mabus believed. It was literally fueling conflict and threatening American lives, he and others found. His premise was that, even leaving aside the security consequences of climate change, reducing reliance on fuel was the best decision for building a lethal and agile military. And whether they are concerned about climate change or not, the one thing everyone in the military can agree on is achieving the mission.

## It's More about Missions than Emissions

The military's own energy transition is not only, or even primarily, about mitigating its contribution to a changing climate but rather about gaining advantage on hostile nations.

"It's more about missions than emissions," said John Conger, former DOD assistant secretary for energy, installations and environment. Larry Farrell Jr., a former air force lieutenant general, couldn't agree more. In our CNA report, he noted the importance of fuel economy on the battlefield, stating, "It's a readiness issue. If you can move your men and materiel more quickly, if you have less tonnage but the same level of protection and firepower, you're more efficient on the battlefield. That's a life and death issue."[11]

Being mission-ready ultimately requires a move away from fossil fuels to a force powered by alternative energies, an effort that began in the 1990s. Following the Kyoto Protocol in 1997, my team at DOD spearheaded the development of DOD's first climate change program.

In December 1998, I delivered a briefing to DOD and White House leaders on the first Department of Defense Emissions Reduction Strategy.[12] My briefing defined the department's energy challenge in two ways: "continue reducing energy use in facilities and in non-tactical vehicles" and "continue developing weapon systems technologies that improve performance while reducing energy use." (In this era, we had not yet fully experienced the perils of sea level rise, persistent wildfires, and extreme weather events that would later compel an equal emphasis on resilience and adaptation.)

These were fine words, but how do we put them into practice within the vast halls of the Pentagon?

The first step was to understand exactly how much fuel the military was using and what the true costs were—not just the market price of oil but the total security costs. Admiral Dick Truly led that effort through a Defense Science Board study I co-commissioned, "More Capable Warfighting through Reduced Fuel Burden" (see chapter 3), in which we found some shocking discrepancies between the price tag for energy and the real expense. The study estimated that in contrast to the $2.31 per gallon the air force was spending on fuel, the actual cost of in-flight refueling was at least $42 per gallon. Admiral Truly and his team recommended standards and certifications that would force DOD to account for the entire cost of getting fuel from the pump to the front, pushing the department to become more energy efficient and justifying investment in alternative energies.[13]

Around the same time, the navy was working to cut energy use on its vessels. Vice Admiral Dennis McGinn, then commander of the US Third Fleet, assembled a group of civilian advisors to help with the effort, including renowned energy expert Amory Lovins. Admiral McGinn had met Lovins through the US Naval War College's Strategic Studies Group and found that he was full of ideas. Lovins and his team at the

Rocky Mountain Institute had already done an evaluation of energy efficiency at the Pentagon building, making recommendations during the five-sided labyrinth's first major renovation since its original construction. Admiral McGinn asked Lovins to do the same for a typical navy ship.

The navy chose the USS *Princeton*, a conventionally powered cruiser. Admiral McGinn described Lovins and his team crawling all over this ship in disguise, "armed with measuring tapes and pressure gauges," to determine how the vessel could "deliver the same energy service with less fuel and uncompromised or improved warfighting capability."[14] Lovins found that improving the electrical and propulsion systems could reduce the ship's total fuel use by 50–75 percent.[15]

Admiral McGinn had high hopes for implementing the recommendations of this study, delivered in late June 2001. But much like the study led by Admiral Truly, also delivered in early 2001 to the incoming George W. Bush administration, it was overtaken by the course of history. The events of September 11 overshadowed all other priorities at the Pentagon.

It would take the mounting deaths of American soldiers trucking fuel to the front in Iraq and Afghanistan to bring attention once more to energy management across the military. Eight of every ten convoys in Iraq and Afghanistan were devoted to supplying fuel and water to the front lines. These trucks were vulnerable to attacks by improvised explosive devices (IEDs), leading to over three thousand US casualties and thousands more injuries, along these transportation routes. This profound tragedy initiated a turning point in the military's approach to fossil fuel use, motivating the highest levels of military leadership to search for alternative ways to power the force. By 2003, General James Mattis of the US Marine Corps, then commanding general of the First Marine Division in Iraq (and later secretary of defense), would call on

DOD to "unleash us from the tether of fuel."[16] His plea, motivated more by improving military capabilities and limiting casualties than by preventing environmental harm, launched a resurgence of efforts to reduce fuel requirements.

This iconic phrase became the guiding vision for yet another Defense Science Board study on DOD's energy strategy, published in 2008. Co-chaired by former secretary of energy and secretary of defense James Schlesinger and former air force vice chief of staff General Michael Carns, the aptly named study, "More Fight—Less Fuel," found that the department had not implemented the recommendations from Admiral Truly's 2001 task force.[17] Tom Morehouse, my first military assistant, who led DOD's early efforts to find alternatives to ozone-depleting halon (see chapter 1), was the primary author of both of these studies.

Morehouse's latest effort, together with numerous reports by the CNA Military Advisory Board and others, provided the analytic foundation for the incoming Obama administration to tackle our forces' fossil fuel dependence. And as is often the case, money spoke louder than words: oil prices at this time were over $140 per barrel, adding heavily to DOD's own fuel costs and motivating President Obama and Secretary Mabus to make energy efficiency a priority within the Pentagon.

Even with these soaring prices, Secretary Mabus's focus on energy surprised some sailors and marines. He set a goal that by 2020, at least half of all navy energy would come from non–fossil fuel sources. "There were all sorts of reasons to do it, but there was a big strategic reason," Secretary Mabus told me. "We were getting our oil from countries that might not have our best interest at heart."

At that point, the military was buying oil from Venezuela, Russia, and Middle Eastern countries. We were refueling our ships in the western Pacific region from a Chinese refinery based in Singapore, Secretary Mabus said. "You really don't want to be dependent on the Chinese warships in the Western Pacific," he said.

Worse, we were losing marines and soldiers in the constant movement of fuel to the front in Afghanistan and Iraq. For every twenty-four fuel convoys in Afghanistan, one soldier was killed; while in Iraq, there was a casualty for every thirty-nine refueling convoys. "That was just way too high a price to pay," Secretary Mabus said to me, his demeanor growing somber.

The urgency of the issue mounted as oil prices spiked once again in 2011 and military planners were forced to make tough choices about whether to pay more or to cut back on their energy consumption. "We were cutting back on steaming, on flying, on training, and that's just not acceptable," Secretary Mabus said. Amid this price increase, he was presented with a $2 billion bill for unbudgeted fuel. Even in the Pentagon, finding $2 billion you had not expected to spend is not an easy task.

"The only way you're going to become completely independent as a country is to produce energy that cannot be exported, that cannot be made, that's immune from the price spikes," he said. "That means producing wind and solar and geothermal. That's producing biofuels and sustainable biofuels for both air and ship propulsion."[18]

What started as purely a warfighting measure to save lives and reduce dependence on foreign sources of oil evolved into an opportunity for the navy to transition to cleaner energy.

By the end of Mabus's eight years as secretary of the navy, two-thirds of energy used across all navy and marine bases came from alternative fuels. His work catalyzed the modern effort to cut our forces' tether to fossil fuel while saving the Pentagon $400 million, which Mabus observed was the price of a littoral combat ship (LCS) warship.

## Cutting the Tether to Fuel

While Secretary Mabus was changing how the navy uses energy, Sharon Burke, assistant secretary of defense for operational energy, was taking up the issue with the secretary of defense. Members of Congress had become increasingly frustrated with DOD's refusal to move faster to reduce its fuel burden, particularly as casualties from trucking fuel to the front were mounting. So Congress established a separate office in the Pentagon to focus on operational energy.

Burke, who had already served at DOD and as a speechwriter at the US Department of State, was tasked with leading the effort, with Tom Morehouse as her principal deputy. Burke brought experience in both DOD and the State Department and was an early leader on environmental matters. Her new office in DOD operational energy had a big job—changing how the department buys and uses energy for military forces—and a small office. But with that office came the authority to push for constructive change and, most importantly, the power of the purse.

In a sign of its seriousness, Congress had granted the office budget certification authority, meaning that Burke and her team reviewed the entire DOD budget and determined whether the department had met its operational energy needs. When Secretary of Defense Ash Carter realized Burke held that power, he asked incredulously, "You don't intend to decertify the president's entire defense budget, do you?"

She certainly did not. "We used that authority more as a scalpel than a hammer to encourage the military departments to meet their goals and to hew to the strategy," Burke told me.[19]

Precise as it was, that scalpel was badly needed. In Afghanistan, the ever-present need for energy was lighting up our positions—in other words, drawing attention to US forces as fuel trucks lumbered over deserts and mountains. "It was no secret where we were, even though we were trying to be low-profile, because we had to move so much fuel in there," Burke said.

Burke's goal was to protect our troops by promoting the work of the army's Rapid Equipping Force, led by Colonel Peter Newell, while ensuring the new equipment was less energy vulnerable for our forces. Sometimes the solutions were remarkably low-tech, like better-insulated tents. At other times they required hybrid or renewable energy technologies such as solar generators. Either way, the energy savings were enormous with just a few basic changes. Sharon Burke's small but mighty team included Rachel Ross, who would later become DOD's deputy chief sustainability officer, and Oliver Fritz, her deputy in the operational energy office, who later went on to direct that office. "What I'm most proud of is that we helped save lives and improve military operations by giving them better energy and better energy choices," Burke said.[20]

At one remote combat post in Afghanistan, soldiers who were firing off mortars were forced to run their vehicles constantly to power the batteries for their mortar guidance systems, Burke said. What if those tanks ran out of gas? Soldiers in combat should not have to worry about whether or not they can operate their equipment or whether they will be caught in a war zone without fuel.

Burke helped to provide the soldiers with a solar-powered generator, and their initial response was absolute shock. "You want me to use *that*?" But the generator proved far more effective and efficient than their old equipment.

"They loved it because, first of all, it worked twenty-four seven because it had a battery on it. Second, it didn't make any noise, whereas the diesel generator they were using before gave off a noise signal that made it obvious where they were. It didn't smell, and they never had problems with making sure they had enough fuel to refill it."[21]

Again, this solar technology was purchased not to save the planet (though cutting emissions is certainly a benefit) but to save lives. "That's how you make hearts and minds of the military, by giving them something that solves a military problem," Burke said.

Energy for the battlefield has traditionally been managed separately from that needed for military bases in the United States. Yet as the nature of warfare evolves with advanced technologies, many modern-day

US military missions, such as operating drones and unmanned vehicles, are performed from a remote computer. Increasingly, the two categories of military fuel consumption, operational energy and installation energy, are becoming more intertwined and critical to DOD's warfighting capabilities.

Today, maintaining power to US bases is "mission-essential," said Joe Bryan, former senior energy and climate advisor to Secretary of Defense Lloyd Austin. Yet, like the rest of us, many of our military installations rely upon commercial utility systems for energy, which means that natural disasters or cyberattacks and physical attacks by our adversaries can turn out the lights.

"What we know is that our adversaries have done a lot of work to understand how to attack our utility systems," Bryan said. It's easy to understand why. "If you want to keep us from being able to deploy or being able to do our jobs, attacking our energy systems is obviously a way to do it."[22]

By building distributed independent electrical grids and smart power systems, our forces can remain online even when the commercial grid is unstable. And they can ensure that mission-critical equipment, including computers and defense systems, stay up and running even when the lights in the gym go out. This approach, sometimes called "islanding the base," is a key part of DOD's climate change strategy, with plans to integrate a microgrid on every installation by 2035.[23]

Rachel Jacobson, assistant secretary of the army (installations, energy and environment), is a key player in that strategy. In June 2022, she attended a ribbon-cutting ceremony at Fort Liberty (formerly Fort Bragg) for its magnificent floating solar array on Big Muddy Lake in North Carolina. Capable of withstanding a Category 5 hurricane, the solar panels provide carbon-free energy generation for the installation; supplement power to the local grid; and support Camp Mackall, a critical section of Fort Liberty, during electrical outages.[24]

This project, at the largest army base in the country, was the first of its kind—constructed in partnership with Duke Energy and Ameresco—but it will hardly be the last. Jacobson and her boss, Secretary of the Army Christine Wormuth, are leading similar collaborations across the army's many bases. In August 2023, they held a ribbon-cutting ceremony to inaugurate a microgrid at Joint Forces Training Base, Los Alamitos in California, which will enable the installation to power itself without the grid for fourteen days.

Microgrids will also help the military succeed in future contests with a peer competitor, such as China, or in regional conflicts in which energy dependence may be weaponized. Since most military bases overseas are vulnerable to power outages from cyberattacks and natural disasters, taking out the power at a US base overseas would be an easy target in a potential fight with China. On the other hand, Russia readily targeted infrastructure and weaponized energy against Ukraine early in Vladimir Putin's war.

The military can avoid this vulnerability by installing microgrids at its overseas bases, according to research by DOD. These bases "would almost be like a mini-city, operating their own power plants without the help of the host nation," according to Lieutenant Colonel Nathan Olsen of the Office of the Undersecretary of Defense for Research and Engineering.[25] DOD is also planning to deploy nuclear microreactors, the size of a standard shipping container, that could power an entire military base. A pilot program at Eielson Air Force Base in Alaska seeks to bring one online in 2027.[26]

And while mission-readiness is the primary goal of these projects, reducing carbon emissions and slowing climate change are welcome side effects. "We recognize that our soldiers are vulnerable to the effects of climate change and that [reducing energy dependence] helps us to be a stronger force," Jacobson told me, "but we also recognize our responsibility to reduce greenhouse gases, our responsibility to manage resilient installations, to be sure we are mitigating the effects of climate change and adapting appropriately so we can be part of the solution."[27]

## A Silver Buckshot

The US military has long played a leading role in developing advanced technologies, from GPS systems to flat-screen televisions to the internet. In each case, the goal has been to maintain the competitive edge for our forces on the battlefield, yet every new technology has had its detractors. As Secretary Mabus told me, "Every time, there were these incredible naysayers: 'That costs money,' or 'You're trading coaling stations all around the world and you're going with this new stuff called oil?' or 'You'll never make nuclear small enough or safe enough to fit inside a submarine.'" But fortunately, that did not stop progress. "And every time, they were wrong. They were wrong again with alternative fuels," Mabus said.[28]

When it comes to energy efficiency, military leaders are determined to be on the right side of history. That is why they are aiming to achieve net-zero energy use by 2050 and to lead the nation in a transition to sustainable fuels. But they can't do it alone. Instead, DOD needs partnerships with the private sector to drive innovation and investment.

A prime example is the effort to electrify DOD's military and nontactical vehicles. As Deputy Secretary of Defense Kathleen Hicks observed, electric vehicles are quiet, are low-maintenance, have incredible torque, and don't require endless gas. All of these attributes "can help give our troops an edge on the battlefield," she said.[29] However, to get this edge, we need the materials to make electric vehicles, which today are mostly found in China, and reliable ways to charge the fleet. Both of these challenges can be solved: the first by diversifying supply chains and the second by ensuring that bases have enough charging stations. Yet to achieve these goals, the military needs help from industry.

Although DOD has more nontactical vehicles than any other department in the federal government, save the US Postal Service, it relies on the automobile industry to electrify its fleet. "While the Department of Defense is a big customer, we're not big enough to drive manufacturing investment back to the United States," Joe Bryan told me. Instead, we need consumer demand and investments by automakers at home to produce electric vehicles at the scale necessary for our forces.

Thus our military's energy transition goes hand in hand with that of our nation as a whole. And collectively, we have the capacity to create profound change.

"We are at a moment in history in which we have amazing opportunities to do new things and use technologies in different ways," Bryan said. These new technologies protect our forces and the civilians they serve and, in the process, reduce climate risks.

Each of our five military branches is heeding the call. At the United States Military Academy West Point, the army created a crosscutting consortium in which cadets learn about all aspects of climate change, how it affects national security, and why the armed services are invested in confronting these threats. Students conclude the program with a capstone project, many of which, Jacobson said in awe, are truly "Nobel Prize worthy." One group of cadets, for instance, created a system for turning waste into energy, and the army plans to pilot it at a base. Another group is creating models of various installations, determining precisely the combination of technologies necessary to achieve 100 percent carbon neutrality.

Likewise, in Albany, Georgia, the Marine Corps has developed a base powered completely by methane from a landfill. And at Fort Meade in Maryland, the army is transforming an empty landfill into a sixty-acre solar farm to power this essential installation.

"Even if you didn't care about climate, you still want to do those things because they're the best decisions you can make to support the military mission," Bryan said.[30]

Building our future force demands a diversity of efforts, a diversity of technologies, and a diversity of energies. As Admiral Dennis McGinn wrote in a 2010 report of the CNA Military Advisory Board, "There is no one perfect new energy solution—we need a silver buckshot approach because there's not a silver bullet."[31]

More than ten years later, the gun with the silver buckshot is beginning to fire.

## A Shared Mission

A decade after concluding his army service, Mike Henchen is proud of the monumental progress the military has achieved in reducing its reliance on fuel and transitioning toward sustainable energy. Things are drastically different from the days when Henchen carefully checked innumerable trucks driving fuel through his post.

One day during his Afghanistan deployment, Henchen met the officer running the small power station at Bagram and inquired about the possibility of integrating solar panels. "He shrugged it off and didn't seem interested at the time," Henchen told me.

Today, countless US installations around the country and the world rely on solar panels and wind turbines to power their missions. For Henchen and the many others who experienced the dangers of our fossil fuel dependence on the ground, this transition profoundly reshapes military service.

Similarly, Robert Hayward, who deployed to Iraq to "turn the lights back on" (see chapter 5), witnessed this generational shift in the military's approach to energy resources over the course of his military service. During his first deployment to Iraq in 2005, the nation's infrastructure relied entirely on fossil fuels. Consequently, a key part of his mission was securing the supply lines that transported oil and gas to power Iraqi facilities—and it was a very dangerous endeavor. "In order to keep these assets, these operations, going, we were putting our soldiers in harm's way by basically kicking them onto the road in order to move these resources across the battlefield," Hayward said.[32]

As top military leaders worked to keep soldiers' lives from being lost while they trucked fuel, Hayward considered this transition from the front in Iraq. "We understood that if you eliminate the fossil fuel aspect of it, you eliminate the immediate security concern," he said. So his transition team started to prioritize the military's movement toward more mobile, lighter, and cleaner forms of energy. It started with hybrid fuels and will eventually move to battery-operated vehicles.

When Hayward returned to Iraq in 2012, the US military was relying more heavily on renewable resources and less on fossil fuels. "We saw the change," he said. "We were moving toward the direction of energy independence."

Energy independence will be critical if we want a key element of our current defense strategy to succeed: namely, deterring conflict with China and ensuring that US forces will prevail should it occur. To win a future war far from the US homeland, the military plans to operate with a "light logistics footprint, using less fuel and dispersed across vast distances," said Deputy Secretary Hicks. "In the Indo-Pacific, it's no stretch to say that operational energy will dictate the margin of victory in a near-peer conflict. Nations that are most resilient and best able to manage the effects of climate change will gain a strategic advantage."[33] To build this strategic advantage and prepare our forces for the battles to come, we must continue moving toward more diverse and dispersed sources of energy.

Today, though no longer in the military, Mike Henchen is working to achieve that goal. After finishing his time in the army, Henchen went on to business school and to a job in the private sector. Yet he found himself longing for the sense of shared mission the military instilled in him, so he made a change. Today, Henchen works for the Rocky Mountain Institute developing strategies to decarbonize the United States' economy through commercial electricity systems and built environments.

"Coming back to a career that feels rooted in mission and purpose, where the people I work with day to day have a shared sense of purpose with me, is motivating," he said. "It's something I missed when I felt I didn't have it."³⁴

The work may feel light-years away from inspecting fuel trucks in Afghanistan, but both come back to one essential factor that affects all our security: energy.

CHAPTER 10

# *Climate-Proofing Security*

IN SEPTEMBER 2023, I traveled from my home in Washington, DC, to The Citadel in Charleston, South Carolina, to commemorate the opening of the school's James B. Near Center for Climate Studies.

One of the nation's oldest military academies, The Citadel is not historically known as a bastion of progressive thought. At the outset of my environmental security career in the 1990s, no one would have guessed such an institution would dedicate an entire program to the study of climate change and its nexus with national security.

The new center will prepare the next generation of military leaders to understand the full breadth of climate impacts shaping our world, emphasizing how climate risks will factor into all military missions. This development builds on a long process of education on the issue, including by the Center for Climate and Security—an organization for which I serve as board chair—which hosted an event at The Citadel on climate security back in 2018.[1]

But for me personally, it had been thirty years since my first official visit to Charleston. In 1993, I traveled with William Perry, then deputy secretary of defense, and Senator Fritz Hollings of South Carolina to close the Charleston Naval Shipyard. On a steamy summer day, Senator Hollings seemed to be the only one who didn't either remove his jacket or break into a sweat. He was likely acclimated to the southeastern heat and humidity, but his pristine attire in a rough shipyard was also a sign of respect and somberness. Loss of the two-hundred-year-old base, critical in the production of navy ships and submarines from World War I through the end of the Cold War, was seen as a major economic blow to the city.

Charleston was one among countless bases that were closed in an effort to resize our military in the aftermath of the Cold War. From the defense buildup of World War II through the beginning of the 1990s, US forces substantially increased their geographic footprint to support more training bases, more shipyards, and more depots. Maintaining this infrastructure was expensive and crowded out investments in modernizing weapons systems, developing new technologies, and training our troops for new missions.

My visit with Secretary Perry and Senator Hollings aimed to support the cleanup of the shipyard while highlighting opportunities across the region to replace the jobs lost. Our role was to try to help Charleston's citizens reimagine their future amid the economic uncertainty of the closure of such a prominent facility.

With the closing of these various military facilities came the responsibility to clean up the contamination of generations' worth of industrial processes, most of which began before the modern era of environmental regulations. (See chapters 2 and 8.) At that time, my work at the US Department of Defense centered on integrating environmental awareness into all aspects of military planning and practices, from protecting wildlife to preventing pollution. We were focused on ensuring that the military was a better environmental steward, but we had yet to imagine all the ways that a changing climate would affect our security.

My first visit to Charleston closed one chapter of the city's military history. My recent visit to The Citadel marked a new one, in which climate change is a core part of the teachings at this historic institution. Today, the military is integrating climate considerations into all aspects of national security. The concept of environmental security has broadened from the military's impact on the environment to the environment's impact on the military. Since climate change is global, influencing every inch of the planet, this is far more complex than simply safeguarding one contained ecosystem or a single base.

Climate change is one of those rare issues that affects every human, every country, every issue, in every possible way. And that includes the military. As US Secretary of Defense Lloyd Austin said, "There is little about what the Department does to defend the American people that is not affected by climate change. It is a national security issue, and we must treat it as such."[2]

Fortunately, today's US military is intentionally preparing to address the compounding crises the climate era presents. Every military branch is accounting for climate risks in its mission, every military academy is including climate change in its curricula, and every DOD service is improving energy resilience and sustainability.

This ambition to understand the complexities our world faces is one of the primary reasons I was drawn to national security in the first place. The military is an institution that requires continuous learning. It is endlessly adapting to new circumstances by developing new technologies, new ways of thinking, and new approaches to best achieve its mission. Although DOD can be slow to change bureaucratic practices—many set by Congress and others—its people, the women and men who serve our country, represent some of the most innovative minds in the nation. This combination of power and purpose is why national security has always excited me. It is not a dead end. It is constantly evolving, never stagnant.

The opening of The Citadel's new climate center perfectly captures our military's ongoing evolution. The center, which overlooks wetlands, is designed to connect with the natural world. As I gazed out at the egrets alighting on marsh grass, I thought of the immense growth of our military and our nation while recognizing the profound challenges that lie ahead. I reflected on the trajectory of my own career—from my earliest days bringing environmental awareness into national security and defense to advising policymakers and others on climate security in my roles with the International Military Council on Climate and Security, the Woodrow Wilson International Center for Scholars, and the Center for Climate and Security. Looking to the future, I offer four central recommendations to transform the threat multiplier of climate change into an opportunity multiplier: awareness, adaptation, mitigation, and alliances.[3] These are the four main pillars of climate action (mitigation and adaptation) and institutional reform (awareness and alliance building). It will take dedication and political will on each front to move the needle toward opportunity.

**Awareness: Improving Climate Prediction**

The first step in understanding the climate threats that lie ahead is to close the gap between short-term weather forecasts and climate models that look decades into the future. This need is not unique to the military. Many sectors of society, from agriculture to shipping to insurance, need better forecasts and modeling; however, when it comes to national security, the approach must be tailored to inform defense decision-making.

Current weather models rely on historical meteorologic records, but unfortunately, forecasts can be outdated if they look only at the past. How do we predict a "hundred-year flood" if events of that magnitude are now happening several times per decade? The answer is deceptively simple: update our models with reliable predictions based on the changing climate, for everything from droughts to hurricanes. While major strides have been made in attributing extreme events to a changing climate, more skill is needed to improve seasonal and subseasonal forecasting. It is easier said than done. Today, we can reliably forecast weather for the next week to ten days, already a remarkable advancement over weather prediction from a few decades ago, which could reliably forecast for only three days. But the further we look into the future, the hazier the picture becomes.

This poses an undeniable problem for our military. Think back to the weather forecasting that played a role in ending World War II: the ability to predict battlefield conditions offered a tremendous tactical advantage. But now, military planners and intelligence analysts are faced with a one- to ten-year gap. This means that defense decision-makers, for both infrastructure and military operations, need to have reliable climate predictions at a local scale for one to ten years in the future. Yet today, we have either reliable weather forecasts for the next week to ten days or climate models that look decades into the future at a global scale.

If we want our military to operate with reliable forecasts, we must close this gap and develop "climate intelligence." One key to improving modeling is to improve systems thinking by examining how the air and sea interact. Since their interplay is a sensitive gauge of changes in the climate system, we must understand both oceanic and atmospheric exchanges rather than examine these factors in isolation. Consider the difference between a baton handoff from one runner to the next and a dance in which partners are constantly coming together and pulling apart. Understanding the dance between ocean and atmosphere is an ever-developing science that allows us to consider climate changes more holistically, and thus to extend short-term weather predictions with greater reliability.

The second component is harnessing today's advanced technologies, including artificial intelligence (AI) and machine learning; digital twins (i.e., digital representations of physical landscapes); and quantum technology, which supercharges sensing, computing, and communications. The advantages of using these technologies range from improving early-warning systems to engaging in better planning with our global partners for natural disasters all over the world. These advancements were unimaginable when I embarked on my national security career, but they are now vital to preparing our military for the challenges ahead.

This brings me to the final aspect of improving climate modeling and weather forecasting: making it relevant for defense planners by integrating natural science and social science. Technology is important, but it does not operate in a vacuum. Without considering how people respond in a crisis, the best models in the world won't prepare us fully for climate disasters and the complexity of their consequences. Migration patterns, power dynamics when communities lack food and water, the impact of extreme storms on livelihoods, and much more all influence the decisions of our forces. To allow the military to plan effectively, models must help us understand human behavior in addition to hard science. That's exactly what DOD is working to do with PREPARES, a program with an apropos acronym for a long title.[4] This program supports research by university social scientists to help military leaders create realistic planning scenarios.

Promising new programs and modeling techniques are being developed every day. But we need to push them forward so that we can do a better job of not only predicting our climate's future but also preparing to meet it—and, where we can, preventing disasters.

## Adaptation: Managing the Unavoidable Impacts of Climate Change

Preparing for climate impacts requires adaptation throughout the military, particularly with regard to three broad areas: education and literacy, military missions, and infrastructure. Caroline Baxter, deputy assistant secretary of defense for force education and training, is leading a department-wide effort to develop climate literacy. "This is not climate change for climate change's sake," she said. "The American way of war is complicated. It is getting more complicated for a myriad of reasons, one of which is that a changing climate is making nearly every aspect of our mission—from strategic issues like maintaining our readiness under any condition to tactical issues like projecting power during extreme weather events—more difficult."[5] For example, we have contingencies for conflicts with China, Russia, North Korea, Iran, and elsewhere, including the deployment of forces—but climate change adds a whole new level of complexity to this preparation. As Leon Panetta, former secretary of defense, told me, we may not be ready for climate-fueled natural disasters that jeopardize our nation and weaken our security. "Are we prepared to deal with major crises of climate," he asked, "where the infrastructure of large parts of the US are shut down and people are left homeless, with need for food, shelter, energy, water, and communication systems?"[6]

Advancing climate literacy also means integrating climate change into the war games our military has played for decades at regional combatant commands around the world. These hypothetical situations help defense decision-makers anticipate and prepare for future scenarios. How will a warming world affect existing war plans, force posture, civilian security, and base resilience? "Climate change has produced a different battlefield for just about every scenario a planner can now imagine, altering the very physical foundations of the geostrategic landscape," Admiral Dave Titley once noted.[7] In response, DOD has begun to include drought, sea level rise, floods, typhoons, and other climate trends in these scenarios. This means that every military decision, from determining how to power a base to planning for a possible ground war in China, must consider the impact of climate change.

Along with this effort, DOD should build a cadre of "climate translators": multidisciplinary specialists who can integrate climate science and modeling with risk assessment and military planning to ensure climate threats are adequately considered across the board. These translators will walk the tightrope between various disciplinary fields and help move our military toward climate resilience.

Climate literacy needs to start early. Military training and academic institutions should mainstream climate change as a core part of their curricula, building a future workforce that can manage the cascading risks of our changing environment. This is already happening at places such as the United States Military Academy West Point and The Citadel's new climate center, and it is spreading to all military institutions. Conventional wisdom on military strategies is no longer sufficient.

The second dimension of adaptation is planning for more humanitarian assistance and disaster relief missions abroad and more support for civilians at home. Historically, "low-intensity conflict," including disaster recovery and counterterrorism measures, had not been the US military's priority. After the attacks of September 11, 2001, terrorism became the defining threat for more than a decade, but strategic competition recently reemerged with Russia's aggression on Ukraine and with China's growing military and political influence. And since October 7, 2023, a new war in the Middle East, in which Iran has been a central strategic player, means that the United States is operating at a high operational tempo ("optempo") around the world.

Today, however, our forces have to "walk and chew gum": be prepared not just for strategic competition and terrorism but also for domestic and international wildfires, floods, hurricanes, earthquakes, and relief operations. We must anticipate multiple "low-intensity disasters," which continuously strain our capabilities, according to Roger Pulwarty, senior scientist with the National Oceanic and Atmospheric Administration.[8] For example, it's not just that hurricanes are more devastating today; it's also that they occur in areas that have already experienced prolonged drought, so the soil is more likely to erode or the forest to catch fire. To have the capacity for these growing missions, our military needs to devote more personnel and resources to search and rescue, lift capacity, firefighting, medical evacuation and support, and other areas that are not traditionally prioritized.

The third dimension is climate-proofing military infrastructure.[9] As we saw in chapter 8, many military bases are highly susceptible to sea level rise, storm surges, flooding, wildfires, permafrost thaw, and more. Today, DOD is turning some of its most vulnerable installations, such as Tyndall Air Force Base in Florida, into climate-resilient "bases of the future." The deadly storm that struck Tyndall in 2018 damaged or destroyed 484 buildings, including hangars for the F-22 fighter jet. Today, Tyndall is one of the nation's largest military construction project, designed to withstand 165 mile per hour winds and seven feet of sea level rise.[10] Tyndall may be the first fully climate-resilient base, but it won't be the last.

Finally, too often, unreasonable barriers stand in the way of technologies that could help transform our energy systems and safeguard our environmental security. Unfortunately, these obstacles are nothing new. In the 1990s, the military bases I visited urgently needed a better way to treat chlorinated solvents in groundwater, but until a new method was certified as "proven," it could not be used. This requires many different regulators in various parts of government to give the okay—lengthening the road new technologies must travel through the proverbial "valley of death" that separates an idea from widespread adoption.

That road is often shorter for weapons systems because militaries are always trying to gain an advantage on the battlefield. When it comes to environmental technologies, however, we do not act with the same urgency to move research from the laboratory to the field.

To get over this hurdle, I cocreated, along with DOD's director of defense research and engineering, Anita Jones, a demonstration and validation program specifically dedicated to environmental technologies used in defense. Although it has a cumbersome name, the Environmental

Security Technology Certification Program (ESTCP) is designed to fast-track promising new research. Starting with a modest DOD budget of $2 million in funding, ESTCP and its companion program, the Strategic Environmental Research and Development Program (SERDP), have today grown to over $200 million.

While these programs have moved forward technologies including treatments for per- and polyfluoroalkyl substances (PFAS), known as "forever chemicals," and the use of fluorine-free firefighting foams, which reduce ozone depletion, there is still more to be done on the climate front. I recommend that DOD create a new, parallel program that will help military leaders understand climate risks in military operations, bases, and communities to build resilience and fast-track vital climate research.

**Mitigation: Creating a Net-Zero Military**

In the twenty-first century, building a resilient military also means changing the way energy is used and produced. The nations that are most energy efficient will not only reduce emissions but also gain a strategic advantage in combat power and military capability.

Most of the energy the military uses is in the form of liquid fuel for aircraft and ships, but new designs are making these vessels more energy efficient—allowing them to reduce their energy consumption while improving their performance. Beyond new designs, the military is looking at alternative fuels. The idea got attention in 2009 when Secretary of the Navy Ray Mabus launched an initiative called the Great Green Fleet, which used blended fuels (a combination of traditional and cleaner alternative fuels) in navy ships and planes. Secretary Mabus said the fleet would "signal to the world America's continued naval supremacy, unleashed from the tether of foreign oil."[11] At the time, biofuels were more expensive than conventional sources, but the math has since begun to shift and those costs are starting to come down.

The price at the pump, stressed Meredith Berger, assistant secretary of the navy (energy, installations, and environment), is something we should never discount. "It's fewer taxpayer dollars. It's less of the budget that's going toward energy because of the efficiency that is provided," she told me.[12]

Energy efficiency is improving not only in our ships and aircraft but also at our bases—a major win in the effort to decarbonize our defense. A good example is Marine Corps Logistics Base Albany in Georgia, which in 2022 became the first installation in the entire department to achieve net-zero energy. Through myriad climate-friendly solutions, including methane capture from landfills, this base now generates more energy than it consumes, reported Secretary of the Navy Carlos Del Toro.[13] This not only cuts emissions but also allows the base to keep the lights on during extreme storms, such as the EF3 tornado that struck the installation in 2017. And, again, it saves the taxpayers money.

The fact that the military is adopting alternative fuels also drives down the costs for everyone else. When DOD, with its substantial budgets, starts purchasing a technology—such as blended fuels or electric vehicles—it affects commercial markets more generally. To multiply the impact of its buying power, DOD is collaborating with the First Movers Coalition, a platform launched by the World Economic Forum at the 2021 United Nations Climate Change Conference (COP26) in Glasgow, Scotland. The premise is to create early markets for innovative clean energy technologies, especially in industries such as shipping, aviation, and trucking that are hard to decarbonize.

Together, DOD leaders are using these myriad efforts to get the department to net-zero emissions by 2050—a goal that would have seemed impossible at the beginning of my career. Today, achieving a net-zero military not only is possible but is within view. As we continue to push toward this goal, we must expand our work with allies and partners, for climate change necessitates global action.

## Alliances: Reimagining Global Cooperation and Competition

The world has changed dramatically since my first days at DOD, but what has not changed, and in fact is truer than ever, is that America's allies are vital to our global security. Case in point: the North Atlantic Treaty Organization's renewed resolve in the face of Vladimir Putin's unprovoked war in Ukraine. In response to Putin's aggression, NATO has added two new members, Finland and Sweden, in a clear indication of democratic unity against authoritarianism.

The hard reality of today's polarized world, however, is that the West's adversaries, from China to Russia to Iran, are becoming more active in regions where conflict is exacerbated by climate change, such as Africa, Latin America, and the Indo-Pacific region. At the same time, global efforts to combat climate change through the United Nations' Conference of the Parties (COP) process are highly uncertain and are widely considered insufficient, according to the US government's own analysis.[14]

Confronted with these dual threats, the military can take several constructive steps to enhance collaborative efforts: sustain a strong presence in regions plagued by climate-fueled conflict; collaborate with other US and international agencies to prepare for natural disasters before they happen; coordinate with allies to make fighting climate change a holistic, global effort; and, finally, support vulnerable nations while countering our competitors' influence.

First, there is no substitute for presence. Whether through port visits or new embassies in remote locations, our diplomacy, development, and defense strategies should all recognize the value of showing up, both in a whole-of-government way and with private sector and civil society partners. Particularly in developing countries, we must truly "lean in" to listen to communities that are vulnerable to climate change. That is how we build trust.

In the 1990s, we developed several programs, such as the Defense

Environmental International Cooperation (DEIC) program and the Arctic Military Environmental Cooperation (AMEC) program, to strengthen alliances by sharing best practices. In recent years, DEIC, now called Defense Operational Resilience International Cooperation (DORIC), has conducted tabletop exercises on reducing disaster risk with partners that are particularly vulnerable to climate change.[15] In addition to preparing for natural disasters, cooperation programs can help promote clean energy and other strategies to reduce emissions. They should be tailored to the specific needs of each region.

In the Arctic, numerous forums for cooperation exist, and most continue to operate even without Russia at the table as a result of its war in Ukraine. In 2021, for instance, DOD launched the Ted Stevens Center for Arctic Security Studies, which brings US and foreign militaries together to address Arctic environmental security. And in 2023, the United States followed in the footsteps of our allies by establishing, for the first time, the position of Arctic ambassador to promote regional diplomacy.

In the Middle East and North Africa, we can promote stability by supporting local efforts to conserve water and address the effects of climate change, as demonstrated by through EcoPeace Middle East. Even allies that are already leaders in clean energy, such as Kenya and Botswana, can benefit from information sharing through events like the Africa Climate Summit 2023 and through joint training and exercises conducted by the US Africa Command.

In the Indo-Pacific region, combating China's growing influence demands more than increasing the number of American troops, ships, and aircraft in the region. In addition, we must strengthen our own regional alliances by supporting work of particular concern to our allies, such as improving early-warning systems for natural disasters and climate-resilient infrastructure. When Secretary of the Navy Del Toro visited the geostrategically significant Pacific Islands in 2022, he reached out to a dozen nations in the "Blue Pacific" to engage them on their "existential climate threats."[16] His visit took place shortly after the Chinese foreign minister led a charm offensive to entice the Pacific Islanders out of the US security orbit. While China is illegally building artificial islands in the South China Sea and militarizing them,[17] the US Department of Defense is working in partnership with nations in the region to build much-needed climate-resilient infrastructure.

Further, in the Caribbean and Latin America, the US military is using drones and other unmanned systems to help regional allies counter their biggest threats at sea, from illegal fishing and narcotics smuggling to human trafficking and natural disasters.

While specific strategies should be tailored to each region, some approaches are applicable anywhere in the world. The second way for our military to collaborate with allies, along with other agencies within our own government, is to collectively plan for natural disasters before they happen. As mentioned earlier, the military is spending more and more time and resources fighting fires, dealing with floods, and providing humanitarian relief. In August 2023 alone, the Center for Climate and Security's Military Responses to Climate Hazards (MiRCH) Tracker identified thirty-five incidents across nineteen countries in which militaries were deployed in response to climate hazards, and that is likely a conservative number.

This trend is only going to increase.

To anticipate these mounting crises, the United States should work with its partners to improve regional military training, planning, and mobilization for humanitarian emergencies. Better preparation means our nation and our allies will be less vulnerable when disaster strikes—and no one will be forced to rely on adversaries or malign actors in times of need.

Third, to be effective, climate collaborations need to be holistic rather than fragmented. Too often, various agencies, with their own authorities and independent funding, conduct global outreach programs without coordinating with domestic and international partners. When it comes to climate security, however, a whole-of-government approach is absolutely vital. Whether you work in the US Department of Defense, Department of State, or Department of Energy, or in the private or nonprofit sector, your climate work must sync up with the work of colleagues at home and abroad. This is a global problem, and it will take a fully coordinated approach to tackle it.

To help unify international climate action, DOD is sharing with allies its DOD Climate Assessment Tool (DCAT), which is used to analyze climate risks at military bases.[18] The agency is also developing the Climate Assessment Tool (CAT), which is not limited to the military but can be used by the United States and our international partners more broadly.[19] These and other modeling systems improve climate readiness around the world, which in turn makes the United States more secure. In Latin America, for example, the US Global Water Security Center shares seasonal forecasts for rain, flooding, and temperatures. This enables both the US Southern Command and our Latin American allies and partners to better plan for drought that could lead to food insecurity in Peru, or heavy rains and floods that could impede delivery of humanitarian aid to embattled Haitians.[20]

Another strategy for global climate action is to create new or build on existing multilateral partnerships, such as the Quadrilateral Security Dialogue, or Quad, a collaboration between four countries—the United States, Japan, Australia, and India—that all have a strategic interest in the Indo-Pacific region. The Quad has recently benefited from Australia's new commitment to climate since its Labor Party came to power. Climate change was discussed extensively at the Quad Leaders' Summit in Tokyo in 2022, where the Quad countries established a climate working group to cut emissions in the Indo-Pacific region and to help neighboring countries prepare and adapt.[21]

The final element of global collaboration on climate change is promoting resilience and prosperity for allies and, in turn, countering the influence of our competitors. There is no question that China and Russia are trying to buy the allegiance of smaller, less affluent nations, not only in the Pacific region but also in Africa, the Middle East, and Latin America. China wants to control much of the world's fishing grounds, forests, farmland, and other natural resources, so it offers seemingly irresistible deals to other countries, such as the construction of critical infrastructure, in exchange for the right to mine, fish, or farm in their territory.

Unless we want the Chinese government to gobble up the planet's critical materials needed for the clean energy transition, we need to offer imperiled nations an alternative. This means putting our money where our mouth is through climate finance for developing countries, something we've been calling for at the Center for Climate and Security for over a decade.[22] The United States should both step up its own climate financing commitments and pressure China to do the same. When US Secretary of the Treasury Janet Yellen went to China in July 2023, she encouraged Beijing to increase its climate spending, sending a clear message to developing countries that helping allies build resilience is a US priority.[23]

The United States can also counter China's influence by building its relationships with small states that are in existential peril from climate change. For instance, the United States inked a new defense and maritime surveillance pact with Papua New Guinea in 2023 and announced the opening of an embassy in Vanuatu—both nations that are threatened by sea level rise, stronger typhoons, and loss of fresh water.[24] These efforts need not be limited to the Pacific but can be replicated in other regions as well.

The growing threat of climate change requires a global all-hands-on-deck approach. That is why international cooperation is critical and why our forces are collaborating with their counterparts around the world to both mitigate climate risks and prepare for the inevitable shocks ahead.

**Toward a Climate-Resilient Future**

Reflecting on how the military has transformed itself from an environmental laggard to a climate and clean energy leader over the past several decades, I am reminded of a quip by Madeleine Albright, former secretary of state: "I'm an optimist who worries a lot."

We will not win every battle against climate change, but it is imperative that we win the war—for ourselves and the next generation. Fortunately, the United States has the strongest, most capable military in the world, and today our forces understand that climate change is a core security concern, shaping every mission. What's more, they are leading by example, not only by combating immediate threats but also by reimagining the ways in which we use energy and preparing for compounding risks.

Over the course of my career, I have had the great privilege of working with those who defend our nation every day, both military and civilian. They have always inspired me to be better, to work harder, and to strive for a higher purpose. Those who serve, along with their families, continue to motivate me to seek and support a more sustainable world.

Despite my deep reverence for our armed forces, I would never have imagined, when I walked into the Pentagon more than thirty years ago, that the military would become the environmental leader it is today. I shouldn't be surprised. Resilience and adaptation have been core military values from time immemorial. The truth is, our military is uniquely equipped to adjust to the rapidly changing conditions triggered by climate change.

Yet global security is not solely a military mission. Whether or not we meet the climate challenge will hinge on whether we, as a people, can come together on this issue, or whether it will continue to divide us. And that is why the military's transformation should inspire us. If generals, charged with the tremendous responsibility of fighting wars, can expand their mission to fight climate change, then the rest of us can rise to the occasion. Rather than remain stymied by ideological differences, we must recognize our common vulnerability and transform this unprecedented threat into an opportunity—an opportunity for prosperity, creativity, innovation, and shared purpose.

# Acknowledgments

My journey to write this book owes much to many. My heroes and heroines are chronicled in this book. From them I have learned a lifetime of lessons about leadership. In my earlier years, I had the privilege of working for Senator Sam Nunn, whose strategic vision for global security and national defense was matched by his astute legislative acumen. My introduction to Senator Nunn was made possible by Robert Murray, one of my earliest mentors at Harvard Kennedy School. I worked with and for Bob Murray during much of my career, including at CNA. Bob allowed my own vision on climate and national security to flourish, for which I am forever grateful. Arnold Punaro, staff director of the US Senate Committee on Armed Services during my time there was also fundamental in supporting my work and was generous with his advice on this book. I am grateful to all my colleagues on the committee.

My many colleagues at the US Department of Defense over the years have inspired me to tell our collective story of mainstreaming the environment, climate, and energy into global security and national defense. My DOD leadership, Secretary William Perry, Secretary Les Aspin, and

Secretary William Cohen, and my direct bosses, Undersecretaries John Deutch, Paul Kaminski, and Jacques "Jack" Gansler, provided the support so essential to move the needle in the Pentagon. Some of my DOD colleagues from the 1990s have helped to reconstruct events, including Jeff Marqusee, Tom Morehouse, Maureen Sullivan, Becky Patton, and Holly Kaufman. There are many more whose military service has been an inspiration for this story: Tracey Walker, Joe Sikes, Kurt Kratz, April Heinze, Kevin Doxey, Roy Solomon, Chris Weaver, and Dave Peters. Carole Parker read early versions of this book and offered valuable insights. Rachel Fleishman has been an intellectual partner from our earliest days at DOD together (she was my first Presidential Management Fellow) to today.

The generals and admirals who served on the CNA Military Advisory Board have been my North Star. We studied the science of climate change together and applied our national security insights to lay the foundation for the field of climate security. I could not have done this work without General Gordon R. Sullivan, the first chair of the CNA Military Advisory Board. His leadership and friendship inspire me every day. Each of the generals and admirals I interviewed for this book were generous with their time and wisdom, including Admiral Skip Bowman, Lieutenant General Larry Farrell Jr., Vice Admiral Paul Gaffney II, Brigadier General Gerry Galloway Jr., Vice Admiral Lee Gunn, General Paul Kern, General Ron Keys, Admiral Joe Lopez, Vice Admiral Denny McGinn, Admiral Joseph Prueher, Rear Admiral David Titley, Vice Admiral Richard Truly, General Chuck Wald, and General Anthony Zinni. Other military leaders whose stories I've been privileged to tell include General Wes Clark, Admiral James Foggo, Admiral Sam Locklear, Rear Admiral Ann Phillips, General Laura Richardson, Admiral Gary Roughead, and Admiral James Stavridis. Secretaries of Defense

William Cohen, Chuck Hagel, Leon Panetta, and William Perry; Deputy Secretary John Deutch; and Secretary of the Navy Ray Mabus have all been gracious with their time and insight.

This book is not only about American leadership on climate security but also about leaders around the world whose work has been an inspiration for this field. Some of those military and national security leaders who have been generous with their time and advice include Brigadier General Luca Baione of Italy, General Munir Muniruzzaman and Shafqat Munir of Bangladesh, General Tom Middendorp of the Netherlands, and Lieutenant General Richard Nugee of the United Kingdom. Others who helped with their stories and advice include Meredith Berger, Annalise Blumm, Steve Brock, Gidon Bromberg, Joe Bryan, Sharon Burke, John Conger, Richard Crusan, Iris Ferguson, Tom Harvey, Robert Hayward, Mike Henchen, Alice Hill, Rachel Jacobson, Mark Nevitt, Elmer Roman, Max Romero, Rachel Ross, Erin Sikorsky, and Swathi Veeravalli.

My colleagues at CNA were "present at the creation" of climate change and national security. I am grateful to my CNA shipmates: David Catarious Jr., Ron Filadelfo, Christine Fox, Henry Gaffney, Noel Gerson, Leo Goff, Marcus King, William Kratz, Sean Maybee, Tom Morehouse, and Kevin Sweeney. Cheryl Rosenblum and Katherine McGrady carry the work forward today. Lee Wasserman (no relation) of the Rockefeller Family Fund made the original project possible through his vision of connecting climate and security.

Greg Douquet has been my intellectual partner in much of this journey. I owe him much for his insights on how marines confront climate and energy in a war zone. Kathy Roth-Douquet has given me astute advice, from the earliest conception of this book to the very end of the process. There was no issue she wouldn't discuss with me on our many dog walks! Ladeene Freimuth has also provided outstanding advice and counsel on many matters throughout our climate and clean energy journey. Chloe Kamarck provided advice on an early book concept. Alice Hill at the Council on Foreign Relations offered great advice on writing, from intellectual ideas to the practical aspects of becoming an author.

My colleagues at the Center for Climate and Security, an institute of the Council on Strategic Risks, have provided expert advice and review of many drafts. I owe special thanks to Francesco Femia and Caitlin Werrell, the founders of the Center for Climate and Security, for their vision, and to Francesco for his reviews. Bob Barnes, Elsa Barron, Catherine Dill, Deb Gordon, Kate Guy, Brigitte Hugh, Christine Parthemore, and Erin Sikorsky have been enormously helpful throughout this process.

I also owe many thanks to my colleagues at the Woodrow Wilson International Center for Scholars. First, Rob Litwak, who encouraged me to come to the Wilson Center and to write a book. Second, the able team in the Scholars' Office: Kimberly Conner and Lindsey Collins. Third, many other colleagues at the Wilson Center who have been intellectual partners on this journey: Dave Balton, Peter Davies, Jack Durkee, Lauren Herzer Risi, Amanda King, Marisol Maddox, Rebecca Pincus, and Mike Sfraga.

At the Atlantic Council, on whose board I serve, the chief executive officer, Fred Kempe, has always set a high bar for writing and leadership on foreign policy and national security. The outstanding team at the Atlantic Council has excelled on so many fronts, including enabling my ideas to have a broader reach. Peter Engelke at the Atlantic Council has also been a policy partner on climate security, and the Veterans Advanced Energy Project has trained the next generation of veteran energy leaders.

My many research fellows and interns have contributed mightily through both research and review. I could not have completed this book without them. My thanks to Mackenzie Allen, Isabella Caltabiano, Zoe Dutton, Lily Feldman, Mariah Fertek, Else Nye, Vanessa Pinney, and Isabel Scal. Two people deserve even more credit for their enormous contributions: Leah Emanuel has put her stellar creativity into various drafts, prodding me to think outside the box and improve my storytelling. Pauline Baudu has been my intellectual partner and research fellow for the past two years, conducting substantive reviews and adding to drafts. I am deeply grateful for their friendship and dedication.

My editor, Emily Turner, has been a dream to work with. From the outset, she gave me great advice and counsel and helped guide me through the writing process. My agent, Don Fehr, has been an excellent coach and guide to the publishing world.

I could not have finished this book without an incredibly supportive family. My parents were my inspiration to dedicate my life to public service and making the world a better place. As Holocaust refugees, they saw firsthand how cruel the world can be. My three children have been my other inspiration. It is for them and their generation that I tell these stories.

My husband, John, has been my lifelong partner. When we worked together at DOD early in the 1990s, a *Washington Post* columnist quipped that John closed the bases and I cleaned them up! My forever thanks to my husband, who provided emotional support, offered excellent editorial comments, and served as a sounding board. He is my greatest champion.

# Notes

**Introduction**

1. CNA Military Advisory Board, "National Security and the Threat of Climate Change," April 15, 2007, https://www.cna.org/reports/2007/national-security-and-the-threat-of-climate-change.

**Chapter 1: From Weapons to Waste**

1. Sherri L. Wasserman, *The Neutron Bomb Controversy: A Study in Alliance Politics*, Foreign Policy Issues (New York: Praeger, 1983).
2. Sherri Wasserman Goodman, "Legal Dilemmas in the Weapons Acquisition Process: The Procurement of the SSN-688 Attack Submarine," *Yale Law & Policy Review* 6, no. 2 (1988): 398, http://www.jstor.org/stable/40239293; J. Ronald Fox et al., "Defense Acquisition Reform, 1960–2009: An Elusive Goal," U.S. Army Center of Military History, 2011, https://history.defense.gov/Portals/70/Documents/acquisition_pub/CMH_Pub_51-3-1.pdf.
3. Sam Nunn, interview by author, June 11, 2023.
4. Nunn, interview.
5. US Nuclear Regulatory Commission, "Backgrounder on the Three Mile Island Accident," last modified November 15, 2022, https://www.nrc.gov/reading-rm/doc-collections/fact-sheets/3mile-isle.html.
6. US Nuclear Regulatory Commission, "Backgrounder on the Three Mile

Island Accident"; Phys.org, "The Three Mile Island Nuclear Accident 40 Years Ago," March 27, 2019, https://phys.org/news/2019-03-mile-island-nuclear-accident-years.html.
7. US Nuclear Regulatory Commission, "Backgrounder on the Three Mile Island Accident."
8. Stuart Diamond, "9 U.S. Reactors Said to Share Characteristics with One in Ukraine," *New York Times*, May 3, 1986, https://www.nytimes.com/1986/05/03/us/9-us-reactors-said-to-share-characteristics-with-one-in-ukraine.html.
9. Andre Carothers, "Plutonium Politics: The Poisoning of South Carolina," *Southern Changes: The Journal of the Southern Regional Council* 10, no. 4 (1988): 4–5, 8–11.
10. For example, see Matthew L. Wald, "Weapon Reactors Faulted on Safety," *New York Times*, October 29, 1987, https://www.nytimes.com/1987/10/29/us/weapon-reactors-faulted-on-safety.html.
11. William Lanouette, "Tritium and *The Times*: How the Nuclear Weapons-Production Scandal Became a National Story," Research Paper R-1, May 1, 1990, Harvard University John F. Kennedy School of Government, Shorenstein Center on Media, Politics and Public Policy, https://shorensteincenter.org/tritium-and-the-times-nuclear-weapons-production-scandal/, https://shorensteincenter.org/wp-content/uploads/2012/03/r01_lanouette.pdf.
12. Lanouette, "Tritium and *The Times*," 9.
13. National Aeronautics and Space Administration, "The Mercury Astronauts," April 9, 2015, https://www.nasa.gov/image-article/mercury-astronauts-2/.
14. Defense Nuclear Facilities Safety Board, "Our Mission," accessed March 17, 2024, https://www.dnfsb.gov/about/mission.
15. Keith Schneider, "The Nation: Rethinking the Arms Complex," *New York Times*, December 25, 1988, https://www.nytimes.com/1988/12/25/weekinreview/the-nation-rethinking-the-arms-complex.html.
16. Al Gore, "A Global Marshall Plan," chap. 15 in *Earth in the Balance: Forging a New Common Purpose*, 1st ed. (London: Routledge, 2007).
17. US Department of Agriculture, Forest Service, "Confronting the Wildfire Crisis," accessed March 17, 2024, https://www.fs.usda.gov/managing-land/wildfire-crisis.
18. Tom Morehouse, interview by author, December 22, 2022.
19. Morehouse, interview.
20. Morehouse, interview.

## Chapter 2: The Birth of Environmental Security

1. George H. W. Bush, address to the United Nations Conference on Environment and Development in Rio de Janeiro, Brazil, June 12, 1992, in *Public

*Papers of the Presidents of the United States: George H. W. Bush*, 924–26 (Washington, DC: US Government Printing Office, 1992), https://www.govinfo.gov/content/pkg/PPP-1992-book1/html/PPP-1992-book1-doc-pg924-2.htm.
2. Associated Press, "Military Pollution Will Remain a Threat for Years, Doctors Report," *Los Angeles Times*, May 16, 1993, https://www.latimes.com/archives/la-xpm-1993-05-16-mn-35995-story.html.
3. *California Desert Lands: Hearings on H.R. 518 and H.R. 880 before the Subcomm. on National Parks, Forests and Public Lands of the Comm. on Natural Resources*, 103rd Cong. (1993), https://www.govinfo.gov/content/pkg/CHRG-103hhrg72354/pdf/CHRG-103hhrg72354.pdf.
4. Joe Lopez, interview by author, March 6, 2023.
5. UVA Miller Center, "Bill Clinton—Key Events," accessed October 15, 2023, https://millercenter.org/president/bill-clinton/key-events.
6. US General Accounting Office, "Military Bases: Analysis of DOD's Recommendations and Selection Process for Closures and Realignments," April 1993, https://www.gao.gov/assets/nsiad-93-173.pdf.
7. Naval History and Heritage Command, "DN-SC-93-05889 President Clinton at NAS Alameda," photograph, August 13, 1993, https://www.history.navy.mil/our-collections/photography/numerical-list-of-images/nhhc-series/naval-subjects-collection/l38-personnel/l38-16-07-02-president-clinton-at-nas-alameda.html.
8. Paul D. Ott, "Many Share Credit for Plan to Contain Plumes of Pollution," *Mashpee (MA) Enterprise*, July 8, 1994.
9. William A. Owens and Sherri W. Goodman, "Good Soldiers Clean Up Their Messes," letter to the editor, *Wall Street Journal*, July 8, 1994.
10. Kevin Dennehy, "A Toast to Clean Water," *Cape Cod Times*, August 29, 2000, https://www.capecodtimes.com/story/news/2000/08/29/a-toast-to-clean-water/51010602007/.
11. Jack Cheevers, "Toxic Legacy: Cleanup at Edwards Lags 5 Years after Its Placement on Priority List," *Los Angeles Times*, August 7, 1995, https://www.latimes.com/archives/la-xpm-1995-08-07-me-32457-story.html; Richard A. Wegman and Harold G. Bailey Jr., "The Challenge of Cleaning Up Military Wastes When U.S. Bases Are Closed," *Ecology Law Quarterly* 21, no. 4 (1994): 865–945, https://www.jstor.org/stable/24113222; "Focus: Military Toxic Sites in Texas," *Texas Environmental Almanac*, accessed March 17, 2024, http://www.texascenter.org/almanac/militarytoxic.htm.
12. US Department of Defense, Environment, Safety & Occupational Health Network and Information Exchange, "Defense Environmental Response Task Force," https://www.denix.osd.mil/arc/denix-files/sites/6/2023/10/Defense-Environmental-Response-Task-Force.pdf.

13. US Department of Defense, "59 FR 1004—Defense Environmental Response Task Force," *Federal Register* 59, no. 5 (January 7, 1994), https://www.govinfo.gov/app/details/FR-1994-01-07/94-303.
14. US Department of Defense, "61 FR 15787—Meeting of the Defense Environmental Response Task Force," *Federal Register* 61, no. 69 (April 9, 1996), https://www.govinfo.gov/app/details/FR-1996-04-09/96-8690; "'Toxic Triangle' Residents Refuse to Go Quietly on Anniversary of Kelly Closure," Deceleration, July 16, 2023, https://deceleration.news/2013/07/16/toxic-triangle-residents-refuse-to-go-quietly-on-anniversary-of-kelly-closure/.
15. US Department of the Navy, Office of the Chief of Naval Operations, "Establishment of Restoration Advisory Boards (RABs)," memo for distribution, February 9, 1994, https://exwc.navfac.navy.mil/Portals/88/Documents/EXWC/Restoration/er_pdfs/gpr/cno-ev-pol-estab%20rabs-1994-02-09.pdf.
16. "EPA Recognizes Lenny Siegel as the Winner of the 2011 Citizen Excellence in Community Involvement Award," https://semspub.epa.gov/work/HQ/175461.pdf.
17. Ralph Vartabedian, "Decades Later, Closed Military Bases Remain a Toxic Menace," *New York Times*, September 27, 2023, updated October 2, 2023, https://www.nytimes.com/2023/09/27/us/military-base-closure-cleanup.html.
18. William K. Stevens, "Wildlife Finds Odd Sanctuary on Military Bases," *New York Times*, January 2, 1996, https://www.nytimes.com/1996/01/02/science/wildlife-finds-odd-sanctuary-on-military-bases.html.
19. Gordon Sullivan, interview by author, May 28, 2022.
20. US Army, Fort Riley, Conservation Branch, "Fort Riley Conservation Projects," accessed November 28, 2023, https://home.army.mil/riley/index.php/about/dir-staff/dpw/env-div/conservation-branch.
21. Dan Charles, "Wee Woodpecker Halts the Tanks," *New Scientist*, December 14, 1991, https://www.newscientist.com/article/mg13217990-500-wee-woodpecker-halts-the-tanks/.
22. Jay Price, "How the Military Helped Bring Back the Red-Cockaded Woodpecker," NPR, *Weekend Edition Saturday*, February 20, 2021, https://www.npr.org/2021/02/20/969703397/how-the-military-helped-bring-back-the-red-cockaded-woodpecker.
23. Price, "How the Military Helped Bring Back the Red-Cockaded Woodpecker."
24. US Army, "National Training Center Continues Environmental Leadership," June 21, 2012, https://www.army.mil/article/82297/national_training_center_continues_environmental_leadership.
25. *California Desert Lands: Hearings on H.R. 518 and H.R. 880 before the*

Subcomm. on National Parks, Forests and Public Lands of the Comm. on Natural Resources, 103rd Cong. (1993).
26. Sherri W. Goodman, "DOD Initiatives: Environmental Security and the Marshall Plan; A Historical Perspective," *Federal Facilities Environmental Journal* 8, no. 2 (Summer 1997): 143–47, https://doi.org/10.1002/ffej.3330080214.
27. Dale Dempsey, "Base's Cleanup Effort Soars," *Dayton (OH) Daily News*, February 16, 1997.

## Chapter 3: Generals and Admirals Battle Climate Change

1. United Nations Climate Change, Process and Meetings, "What Is the Kyoto Protocol?," accessed August 8, 2023, https://unfccc.int/kyoto_protocol.
2. "DOD Climate Treaty and National Security," memo from author's files, November 1998.
3. *Cambridge Dictionary*, s.v. "Bomb Cyclone," accessed August 2, 2023, https://dictionary.cambridge.org/us/dictionary/english/bomb-cyclone; National Weather Service, "Derecho," accessed August 2, 2023, https://www.weather.gov/lmk/derecho.
4. *S.Res.98—A Resolution Expressing the Sense of the Senate regarding the Conditions for the United States Becoming a Signatory to Any International Agreement on Greenhouse Gas Emissions under the United Nations Framework Convention on Climate Change*, 105th Cong. (1997–1998), https://www.congress.gov/bill/105th-congress/senate-resolution/98.
5. Remarks by Sherri Wasserman Goodman, *National Journal*, October 1997.
6. "Memorandum, National Economic Council Director, Annotated Copy, Eugene B. ('Gene') Sperling [et al.] to President Clinton, Subject: Climate Change Decision Memorandum," National Security Archive, October 18, 1997, https://nsarchive.gwu.edu/document/28137-document-6-memorandum-national-economic-council-director-annotated-copy-eugene-b.
7. Robert Holzer, "Pollutions Treaty Poses Threat to US Military," *Defense News*, October 18, 1997.
8. "Climate Change and the Military: Examining the Pentagon's Integration of National Security Interests and Environmental Goals under Clinton," National Security Archive, May 26, 2022, https://nsarchive.gwu.edu/briefing-book/environmental-diplomacy/2022-05-26/climate-change-and-military-examining-pentagons.
9. Robert Holzer, "Pollution Treaty Poses Threat to US Military," *Defense News*, October 18, 1997.
10. "Climate Change and the Military."
11. "National Security Council Memorandum, Climate Change," March 27, 1998, from author's files.

12. Sherri Goodman, "Department of Defense Climate Change Programs," December 12, 1998, from author's files.
13. Goodman, "Department of Defense Climate Change Programs," December 12, 1998, from author's files.
14. US Department of Defense, Office of the Undersecretary of Defense for Acquisition and Technology, "Report of the Defense Science Board on More Capable Warfighting through Reduced Fuel Burden," May 2001, https://dsb.cto.mil/reports/2000s/ADA392666.pdf.
15. Dick Truly, interview by author, August 2, 2022.
16. US Environmental Protection Agency, "32 Citizens, Groups Honored for Addressing Global Warming, Ozone Depletion," October 31, 2000, https://www.epa.gov/archive/epapages/newsroom_archive/newsreleases/c44b9868412942078525698900734b5d.html.
17. David Eady et al., "Sustain the Mission Project: Casualty Factors for Fuel and Water Resupply Convoys," Army Environmental Policy Institute report, September 2009, https://www.researchgate.net/publication/235140243_Sustain_the_Mission_Project_Casualty_Factors_for_Fuel_and_Water_Resupply_Convoys; Associated Press, "Soldiers Battle New Foe in Iraq: Unbearable Heat," *East Bay (San Francisco Bay Area) Times*, July 21, 2007, https://www.eastbaytimes.com/2007/07/21/soldiers-battle-new-foe-in-iraq-unbearable-heat/.
18. Antoine Soubeyran and Agnes Tomini, "Water Shortages and Conflict," *Revue d'économie politique* 122, no. 2 (2012): 279–97, https://www.cairn.info/revue-d-economie-politique-2012-2-page-279.htm.
19. *An Inconvenient Truth: A Global Warning*, 2006 documentary starring Al Gore, directed by Davis Guggenheim, https://algore.com/library/an-inconvenient-truth-dvd.
20. J. Hansen et al., "Global Climate Change as Forecast by Goddard Institute for Space Studies Three-Dimensional Model," *Journal of Geophysical Research* 93, no. D8 (August 20, 1988): 9341–64, https://pubs.giss.nasa.gov/docs/1988/1988_Hansen_ha02700w.pdf.
21. Terry P. Kelley, "Global Climate Change Implications for the United States Navy," United States Naval War College, May 1990, https://documents.theblackvault.com/documents/weather/climatechange/globalclimatechange-navy.pdf.
22. Center for Climate Integrity, "Lie-brary," accessed August 8, 2023, https://climateintegrity.org/lie-brary?gclid=Cj0KCQjwz8emBhDrARIsANNJjS6YNkeFh4OgoILvR7I9sS6Lcs3EnSQKPat9urXp3UxCWqGE7uzhCsYaAlR

KEALw_wcB; Phoebe Keane, "How the Oil Industry Made Us Doubt Climate Change," BBC News, September 19, 2020, https://www.bbc.com/news/stories-53640382; Gabriel Malek and Andrew Baxter, "Oil and Gas M&A Is Undermining the Energy Transition. It's Time to Act," Environmental Defense Fund, May 10, 2022, https://business.edf.org/insights/oil-and-gas-ma-is-undermining-the-energy-transition-its-time-to-act/.

23. Lee Wasserman, interview by author, February 24, 2023.
24. Stephen M. R. Covey and Rebecca R. Merrill, *The Speed of Trust: The One Thing That Changes Everything* (New York: Free Press, 2006).
25. Truly, interview.
26. "Testimony of General Gordon R. Sullivan, USA (Ret.), before the Select Committee on Energy Independence and Global Warming, U.S. House of Representatives," April 18, 2007, https://www.markey.senate.gov/imo/media/globalwarming/tools/assets/files/0116.pdf.
27. Truly, interview.
28. Rian van der Merwe, "Product Management and the Fog of War," *Elezea* (blog), March 22, 2023, https://elezea.com/2023/03/product-management-and-the-fog-of-war/.
29. CNA Military Advisory Board, "National Security and the Threat of Climate Change," April 15, 2007, https://www.cna.org/reports/2007/national-security-and-the-threat-of-climate-change.
30. Skip Bowman, interview by author, February 24, 2023.
31. Emily Sohn, "Climate Change and the Rise and Fall of Civilizations," National Aeronautics and Space Administration, January 20, 2014, https://climate.nasa.gov/news/1010/climate-change-and-the-rise-and-fall-of-civilizations/.
32. CNA Military Advisory Board, "National Security and the Threat of Climate Change."
33. CNA Military Advisory Board, "National Security and the Threat of Climate Change."
34. Justin Worland, "Why Big Business Is Taking Climate Change Seriously," *TIME*, September 23, 2015, https://time.com/4045572/big-business-climate-change/.
35. CNA Military Advisory Board, "National Security and the Threat of Climate Change."
36. Sherri Goodman and Pauline Baudu, "Briefer: Climate Change as a 'Threat Multiplier': History, Uses and Future of the Concept," Center for Climate and Security, January 3, 2023, https://climateandsecurity.org/2023/01/briefer-climate-change-as-a-threat-multiplier-history-uses-and-future-of-the

-concept/. For the 2007 report, see CNA Military Advisory Board, "National Security and the Threat of Climate Change," cited earlier in this chapter.
37. United Nations, "Security Council Holds First-Ever Debate on Impact of Climate Change on Peace, Security, Hearing over 50 Speakers," press release SC/9000, April 17, 2007, https://press.un.org/en/2007/sc9000.doc.htm.
38. Andrew Clark, "Climate Change Threatens Security, UK Tells UN," *Guardian*, April 18, 2007, https://www.theguardian.com/environment/2007/apr/18/greenpolitics.climatechange.
39. Reuters, "U.N. Council Hits Impasse over Debate on Warming," *New York Times*, April 18, 2007, https://www.nytimes.com/2007/04/18/world/18nations.html.
40. Ed Markey, "April 18, 2007—Geopolitical Implications of Rising Oil Dependence and Global Warming," press release, April 17, 2007, https://www.markey.senate.gov/news/press-releases/2007/04/17/april-18-2007-geopolitical-implications-of-rising-oil-dependence-and-global-warming.
41. *Climate Change: National Security Threats; Hearing before the Comm. on Foreign Relations, United States Senate*, 110th Cong. (2007), https://www.govinfo.gov/content/pkg/CHRG-110shrg42725/html/CHRG-110shrg42725.htm.
42. *S.1018—Global Climate Change Security Oversight Act*, 110th Cong. (2007–2008), https://www.congress.gov/bill/110th-congress/senate-bill/1018; *S.3036—Lieberman-Warner Climate Security Act of 2008*, 110th Cong. (2007–2008), https://www.congress.gov/bill/110th-congress/senate-bill/3036.
43. *H.R.2454—American Clean Energy and Security Act of 2009*, 111th Cong. (2009–2010), https://www.congress.gov/bill/111th-congress/house-bill/2454/text.
44. US Department of Defense, "Quadrennial Defense Review Report," February 2010, 83, https://history.defense.gov/Portals/70/Documents/quadrennial/QDR2010.pdf.
45. Stéphanie Giry, "Climate Conflicts," *New York Times Magazine*, December 9, 2007, https://www.nytimes.com/2007/12/09/magazine/09climateconflict.html.
46. Council on Strategic Risks, "Center for Climate and Security," accessed March 17, 2024, https://councilonstrategicrisks.org/ccs/.
47. Center for Climate and Security, "Policy," accessed November 27, 2023, https://climateandsecurity.org/policy/; Center for Climate and Security, "The Climate and Security Advisory Group (CSAG): A Climate Security Plan for America," September 2019, https://climateandsecurity.org/climatesecurityplanforamerica/.
48. The organization has also produced cutting-edge analysis, including on the

effects of climate change on instability and conflict in Syria. Its analysis has had a major impact on the public discourse and has opened the door for significant scientific research on the issue in the years since. This work by the Center for Climate and Security, inspired by our "threat multiplier" report in 2007 (see CNA Military Advisory Board, "National Security and the Threat of Climate Change," cited earlier in this chapter), has contributed to, as its co-founder Caitlin Werrell once noted, bringing climate change "to the big kid's table" of national and international security (see introductory text to *Climate & Security Podcast*, accessed March 26, 2024, https://climateandsecurity.org/podcast/).
49. Sharon Burke, "Achieving Climate Security," United States Institute of Peace, Discussion Paper 23-006, August 15, 2023, https://www.usip.org/publications/2023/08/achieving-climate-security.
50. John Conger, "Sherri Goodman and Vice Admiral McGinn Testify before House Foreign Affairs Committee," Center for Climate and Security, April 4, 2019, https://climateandsecurity.org/2019/04/sherri-goodman-and-vice-admiral-mcginn-testify-before-house-foreign-affairs-committee/; "Letter to the President of the United States: 58 Senior Military and National Security Leaders Denounce NSC Climate Panel," Center for Climate and Security, March 5, 2019, https://climateandsecurity.org/letter-to-the-president-of-the-united-states-nsc-climate-panel/.
51. Conservatives for Clean Energy, accessed November 27, 2023, https://cleanenergyconservatives.com/; Evangelical Environmental Network, accessed November 27, 2023, https://creationcare.org/.
52. "Testimony of Senator John Warner (Retired), Senate Environment and Public Works Committee, October 28, 2009," https://www.epw.senate.gov/public/_cache/files/c/2/c2d1df21-7242-43c5-81c2-9d856c1a8a6f/01AFD79733D77F24A71FEF9DAFCCB056.testimonyofsenatorjohnwarnerepw102809docx2.pdf.

## Chapter 4: Melting Ice and Rising Tensions in the Arctic

1. James Foggo, interview by author, August 25, 2023.
2. CNA Military Advisory Board, "National Security and the Threat of Climate Change," May 1, 2007, 38, https://www.cna.org/reports/2007/national-security-and-the-threat-of-climate-change.
3. Timothy Clack, Ziya Meral, and Louise Selisny, eds., *Climate Change, Conflict, and (In)Security: Hot War*, 1st ed. (London: Routledge, 2024); Sherri Goodman, Pauline Baudu, and Rachel Fleishman, "Maritime Response to Climate Change," chap. 10 in Clack, Meral, and Selisny, *Climate Change, Conflict, and (In)Security*.

4. US Department of Commerce, National Oceanic and Atmospheric Administration, NOAA in the Arctic, "Arctic Report Card," https://arctic.noaa.gov/report-card/ (for 2006, see https://arctic.noaa.gov/wp-content/uploads/2023/04/ReportCard2006.pdf).
5. Katarina Kertysova and Akash Ramnath, "How Permafrost Thaw Puts the Russian Arctic at Risk," IPI Global Observatory, November 22, 2021, https://theglobalobservatory.org/2021/11/how-permafrost-thaw-puts-the-russian-arctic-at-risk/; Jaroslav Obu et al., "Northern Hemisphere Permafrost Map Based on TTOP Modelling for 2000–2016 at 1 km$^2$ Scale," *Earth-Science Reviews* 193 (June 2019): 299–316, https://doi.org/10.1016/j.earscirev.2019.04.023.
6. US Department of Commerce, National Oceanic and Atmospheric Administration, NOAA in the Arctic, "Arctic Report Card: Update for 2022," 2022, https://arctic.noaa.gov/report-card/report-card-2022/.
7. Robert D. McWethy, "Significance of the *Nautilus* Polar Cruise," *United States Naval Institute Proceedings* 84, no. 5 (May 1958): 663, https://www.usni.org/magazines/proceedings/1958/may/significance-nautilus-polar-cruise.
8. Geoff Ziezulewicz, "In a Thawing Era, ICEX 2020 Kicks Off Up North," *Navy Times*, March 5, 2020, https://www.navytimes.com/news/your-navy/2020/03/05/in-a-thawing-era-icex-2020-kicks-off-up-north/.
9. Caitlin Werrell and Francesco Femia, "Climate Security Q&A with Admiral Titley and Admiral Morisetti," Center for Climate and Security, November 26, 2014, https://climateandsecurity.org/2014/11/climate-security-qa-with-admiral-titley-and-admiral-morisetti/.
10. CNA Military Advisory Board, "National Security and the Threat of Climate Change."
11. Thomas Nilsen, "Sea Ice at the North Pole Was Too Thin to Properly Test Russia's Giant New Icebreaker," *Arctic Business Journal*, October 22, 2020, https://www.arctictoday.com/sea-ice-at-the-north-pole-was-too-thin-to-properly-test-russias-giant-new-icebreaker/.
12. Pauline Baudu, "Minding the Archipelago: What Svalbard Means to NATO," *Arctic Review on Law and Politics* 14 (January 6, 2023): 76–82, https://doi.org/10.23865/arctic.v14.5197.
13. Thomas Nilsen, "Moscow Aims to Enhance Presence in Svalbard as Part of Hybrid-Strategy, Expert Warns," *Barents Observer*, December 7, 2021, https://thebarentsobserver.com/en/security/2021/12/moscow-aims-enhance-presence-svalbard-part-hybrid-strategy-expert-warns.
14. Baudu, "Minding the Archipelago."
15. National Intelligence Council, "National Intelligence Estimate: Climate

Change and International Responses Increasing Challenges to US National Security through 2040," NIC-NIE-2021-10030-A, 2021, 8, https://www.dni.gov/files/ODNI/documents/assessments/NIE_Climate_Change_and_National_Security.pdf.

16. Melody Schreiber, "A Russian Incursion into Ukraine Will Likely Affect the Arctic—but Exactly How Is Unclear," *Nunatsiaq News*, January 27, 2022, https://nunatsiaq.com/stories/article/a-russian-incursion-into-ukraine-will-likely-affect-the-arctic-but-exactly-how-is-unclear/.
17. Sherri Goodman et al., "Inclusive Planning for Changing Arctic Futures: Demonstrating a Scenario-Based Discussion; A Tabletop Exercise Demonstration at the Arctic Futures 2050 Conference," September 5, 2019, https://councilonstrategicrisks.org/wp-content/uploads/2019/09/2050-Arctic-Tabletop-Report.pdf. Special thanks to Jim Townsend.
18. Vladislav Inozemtsev, "Russia's Northern Sea Route Ambitions," *Eurasia Daily Monitor* 13, no. 84 (April 29, 2016), https://jamestown.org/program/russias-northern-sea-route-ambitions/.
19. Sherri Goodman and Elisabeth Freese, "China's Ready to Cash In on a Melting Arctic," *Foreign Policy*, May 1, 2018, https://foreignpolicy.com/2018/05/01/chinas-ready-to-cash-in-on-a-melting-arctic/; Jane Nakano and William Li, "China Launches the Polar Silk Road," Center for Strategic and International Studies, February 2, 2018, https://www.csis.org/analysis/china-launches-polar-silk-road.
20. Thad Allen, interview by author, November 17, 2023.
21. Government of Canada, National Defence, "Third New Arctic and Offshore Patrol Ship Delivered to Canada," news release, September 2, 2022, https://www.canada.ca/en/department-national-defence/news/2022/09/third-new-arctic-and-offshore-patrol-ship-delivered-to-canada.html; Centre for Ocean Applied Sustainable Technologies, "*COAST Talks*: Weathering Change—How Ocean Tech Supports Climate Security," podcast with Sherri Goodman and Angus Topshee, March 15, 2023, https://canadacoast.ca/event/coast-talks-weathering-change-how-ocean-tech-supports-climate-security/.
22. ArcticToday, "Canada's New Naval Arctic Patrol Ship Circumnavigated North America on Its First Voyage," *Arctic Business Journal*, December 17, 2021, https://www.arctictoday.com/canadas-new-naval-arctic-patrol-ship-circumnavigated-north-america-on-its-first-voyage/.
23. David Balton, "The Arctic Fisheries Agreement Enters into Force," Polar Institute, *Polar Points* (blog), June 25, 2021, https://www.wilsoncenter.org/blog-post/no-9-arctic-fisheries-agreement-enters-force.
24. Balton, "Arctic Fisheries Agreement Enters into Force."

25. Hope McKenney and Maggie Nelson, "Sighting of Chinese and Russian Warships near Aleutians Prompts Navy Response," Alaska Public Media, August 8, 2023, https://alaskapublic.org/2023/08/08/sighting-of-chinese-and-russian-warships-near-aleutians-prompts-navy-response/.
26. Steven Lee Myers et al., "How China Targets the Global Fish Supply," *New York Times*, September 26, 2022, https://www.nytimes.com/interactive/2022/09/26/world/asia/china-fishing-south-america.html; Financial Transparency Coalition, "Half of Industrial IUU Fishing Vessels Operate in Africa, Majority Chinese and European—New Report," October 26, 2022, https://financialtransparency.org/half-illegal-fishing-vessels-operate-africa-majority-chinese-european-new-report/.
27. "U.S. 'Climate Weapon' Caused Russia's Warm Winter, Lawmaker Says," *Moscow Times*, January 15, 2020, updated September 7, 2021, https://www.themoscowtimes.com/2020/01/15/us-climate-weapon-caused-russias-warm-winter-lawmaker-says-a68904.
28. The Arctic Seven are the Arctic coastal states that were the original members of the Arctic Council: the United States, Canada, Norway, Finland, Sweden, Iceland, and Denmark.
29. Marisol Maddox, "Climate-Fragility Risk Brief: The Arctic," Climate Security Expert Network, July 2021, https://www.wilsoncenter.org/sites/default/files/media/uploads/documents/csen_risk_brief_arctic.pdf.
30. Sherri Goodman and Katarina Kertysova, "The Nuclearization of the Russian Arctic: New Reactors, New Risks," European Leadership Network, June 26, 2020, https://www.europeanleadershipnetwork.org/policy-brief/the-nuclearisation-of-the-russian-arctic-new-reactors-new-risks/.
31. Alec Luhn, "Russia's 'Slow-Motion Chernobyl' at Sea," BBC, September 1, 2020, https://www.bbc.com/future/article/20200901-the-radioactive-risk-of-sunken-nuclear-soviet-submarines.
32. My deputy, Gary Vest, played an important role in developing the AMEC program. Bob Edson, Dieter Rudolph, and Susan Clark-Sestak also had key roles.
33. Sherri Goodman, "Changing Climates for Arctic Security," *Wilson Quarterly*, Summer 2017, https://www.wilsonquarterly.com/quarterly/into-the-arctic/changing-climates-for-arctic-security.
34. John Deutch, "The Environment on the Intelligence Agenda," World Affairs Council, July 25, 1996, https://www.nro.gov/Portals/65/documents/news/press/1996/1996-06.pdf.
35. Sherri Wasserman Goodman and Sergei Grigorov, "Defense Cleanup: International Cooperation," *Engineering News Record*, March 24, 1997.

## Chapter 5: Drought, Oil, and Power in Africa and the Middle East

1. Robert Hayward, interview by author, May 25, 2023.
2. Hayward, interview.
3. Nikolaos Christidis, Dann Mitchell, and Peter A. Stott, "Rapidly Increasing Likelihood of Exceeding 50 °C in Parts of the Mediterranean and the Middle East Due to Human Influence," *NPJ Climate and Atmospheric Science* 6, article no. 45 (2023), https://doi.org/10.1038/s41612-023-00377-4; Khalil Abu Allan et al., "How Is Climate Change Affecting MENA? Local Experts Weigh In," Woodrow Wilson International Center for Scholars, April 20, 2023, https://www.wilsoncenter.org/article/how-climate-change-affecting-mena-local-experts-weigh; London School of Hygiene & Tropical Medicine, "Limiting Warming to 2°C May Avoid 80% of Heat-Related Deaths in Middle East and North Africa," ScienceDaily, April 4, 2023, https://www.sciencedaily.com/releases/2023/04/230404114158.htm.
4. Frederic Wehrey and Ninar Fawal, "Cascading Climate Effects in the Middle East and North Africa: Adapting through Inclusive Governance," Carnegie Endowment for International Peace, February 24, 2022, https://carnegieendowment.org/2022/02/24/cascading-climate-effects-in-middle-east-and-north-africa-adapting-through-inclusive-governance-pub-86510; United Nations Development Programme, "Dynamics of Violent Extremism in Africa: Conflict Ecosystems, Political Ecology and the Spread of the Proto-State," February 7, 2023, https://www.undp.org/africa/publications/dynamics-violent-extremism-africa-conflict-ecosystems-political-ecology-and-spread-proto-state.
5. United Nations Framework Convention on Climate Change, "COP28 Agreement Signals 'Beginning of the End' of the Fossil Fuel Era," December 13, 2023, https://unfccc.int/news/cop28-agreement-signals-beginning-of-the-end-of-the-fossil-fuel-era.
6. Matt Ince and Erin Sikorsky, "The Uncomfortable Geopolitics of the Clean Energy Transition," *Lawfare*, December 13, 2023, https://www.lawfaremedia.org/article/the-uncomfortable-geopolitics-of-the-clean-energy-transition.
7. Anthony Zinni, interview by author, August 2, 2022.
8. Zinni, interview.
9. Zinni, interview.
10. Brigadier General Joseph G. Garrett III, U.S. Army, "The Army and the Environment: Environmental Considerations during Army Operations," *International Law Studies* 69 (1996), https://digital-commons.usnwc.edu/cgi/viewcontent.cgi?article=1539&context=ils.

11. International Military Council on Climate and Security, https://imccs.org/; Tom Middendorp, interview by author, April 25, 2023.
12. Middendorp, interview.
13. Middendorp, interview.
14. Statista, "The 20 Countries with the Highest Fertility Rates in 2023," last modified September 27, 2023, https://www.statista.com/statistics/262884/countries-with-the-highest-fertility-rates/.
15. CNA Military Advisory Board, "The Role of Water Stress in Instability and Conflict," December 2017, https://www.cna.org/reports/2017/CRM-2017-U-016532-Final.pdf.
16. Marcus D. King, *Weaponizing Water: Water Stress and Islamic Extremist Violence in Africa and the Middle East* (Boulder, CO: Lynne Rienner, 2023).
17. Francesco Femia and Caitlin E. Werrell, "Syria: Climate Change, Drought and Social Unrest," Center for Climate and Security, Briefer no. 11, February 29, 2012, https://climateandsecurity.files.wordpress.com/2012/04/syria-climate-change-drought-and-social-unrest_briefer-11.pdf.
18. Climate Signals, "Syrian Drought 2007–2010," last updated October 15, 2021, https://www.climatesignals.org/events/syrian-drought-2007-2010; Colin P. Kelley et al., "Climate Change in the Fertile Crescent and Implications of the Recent Syrian Drought," *Earth, Atmospheric, and Planetary Sciences* 112, no. 11 (March 2, 2015): 3241–46, https://doi.org/10.1073/pnas.1421533112.
19. Caitlin E. Werrell, Francesco Femia, and Troy Sternberg, "Did We See It Coming? State Fragility, Climate Vulnerability, and the Uprisings in Syria and Egypt," *SAIS Review of International Affairs* 35, no. 1 (Winter–Spring 2015): 29–46, https://doi.org/10.1353/sais.2015.0002.
20. Office of the United Nations High Commissioner for Human Rights, Special Procedures of the Human Rights Council, "UN Special Rapporteur on the Right to Food: Mission to Syria from 29 August to 7 September 2010," September 7, 2010, https://www.ohchr.org/sites/default/files/english/issues/food/docs/SyriaMissionPreliminaryConclusions_07092010.pdf.
21. King, *Weaponizing Water*.
22. Marcus D. King and Julia Burnell, "The Weaponization of Water in a Changing Climate," in *Epicenters of Climate and Security: The New Geostrategic Landscape of the Anthropocene*, edited by Caitlin E. Werrell and Francesco Femia, 67–73 (Washington, DC: Center for Climate and Security, June 2017), https://climateandsecurity.org/wp-content/uploads/2017/06/8_water-weaponization.pdf.
23. King, *Weaponizing Water*.
24. King, *Weaponizing Water*.

25. Gideon Bromberg, interview by author, May 1, 2023.
26. Tania Krämer, "Israel and Jordan's Climate Deal," Deutsche Welle (DW), January 19, 2022, https://www.dw.com/en/former-foes-israel-and-jordan-work-together-to-combat-energy-and-water-scarcity/a-60310997.
27. Ram Aviram, Ahmad Hindi, and Saad Abu Hammour, "Coping with Water Scarcity in the Jordan River Basin," Century Foundation, December 14, 2020, https://tcf.org/content/report/coping-water-scarcity-jordan-river-basin/; Global Nature Fund, "Dead Sea—Israel, Jordan, Palestine," accessed November 25, 2023, https://www.globalnature.org/en/living-lakes/asia/the-dead-sea.
28. Oded Balilty and Ilan Ben Zion, "AP Photos: Lingering Drought Threatens Holy Land's Waters," AP, December 25, 2017, https://apnews.com/article/dbaa115be96642cf90530b331ba31f44.
29. "Jordan says it won't sign energy for water deal with Israel," *Reuters*, November 16, 2023, https://www.reuters.com/world/middle-east/jordan-says-it-wont-sign-energy-water-deal-with-israel-2023-11-16/
30. Leah Emanuel, "Collaborating across Borders: Young Professionals in the Middle East Tackle Region's Water Issues," Woodrow Wilson International Center for Scholars, Environmental Change and Security Program, *New Security Beat* (blog), November 3, 2020, https://www.newsecuritybeat.org/2020/11/collaborating-borders-young-professionals-middle-east-tackle-regions-water-issues/.
31. Emanuel, "Collaborating across Borders."
32. Swathi Veeravalli, interview by author, June 2, 2023.
33. Veeravalli, interview.
34. Veeravalli, interview.
35. Veeravalli, interview.

## Chapter 6: Navigating Asia's Disaster Alley

1. Lee Gunn, interview by author, March 8, 2023.
2. Joseph Nye, *Soft Power: The Means to Success in World Politics* (New York: Public Affairs, 2004).
3. Gunn, interview.
4. Samuel Locklear, interview by author, March 20, 2023.
5. Locklear, interview.
6. Samuel J. Locklear III, foreword to "The U.S. Asia-Pacific Rebalance, National Security and Climate Change," edited by Caitlin E. Werrell and Francesco Femia, Center for Climate and Security, November 2015, 9, https://climateandsecurity.org/wp-content/uploads/2015/11/ccs_us_asia_pacific-rebalance_national-security-and-climate-change.pdf.

7. Princeton University, Nuclear Princeton, "The Pacific: Atomic Bomb Testing at Bikini Atoll 1946," accessed August 8, 2023, https://nuclearprinceton.princeton.edu/pacific.
8. Atomic Heritage Foundation, "Marshall Islands," accessed August 8, 2023, https://ahf.nuclearmuseum.org/ahf/location/marshall-islands/.
9. CNA Military Advisory Board, "National Security and the Threat of Climate Change," May 1, 2007, 25, https://www.cna.org/reports/2007/national-security-and-the-threat-of-climate-change.
10. CNA Corporation, "Climate Change Poses Serious Threat to U.S. National Security," ScienceDaily, April 17, 2007, https://www.sciencedaily.com/releases/2007/04/070417092232.htm.
11. "Testimony of Admiral Joseph Prueher, USN (Ret.), before the Committee on Foreign Relations, U.S. Senate, May 9, 2007," https://www.foreign.senate.gov/imo/media/doc/PrueherTestimony070509.pdf.
12. "Testimony of Admiral Joseph Prueher," 3.
13. "Statement of Admiral Samuel J. Locklear, U.S. Navy Commander, U.S. Pacific Command, before the Senate Armed Services Committee on U.S. Pacific Command Posture, 9 April 2013," https://www.armed-services.senate.gov/imo/media/doc/Locklear-04-09-13.pdf; Caitlin Werrell and Francesco Femia, "Admiral Locklear: Climate Change the Biggest Long-Term Security Threat in the Pacific Region," Center for Climate and Security, March 12, 2013, https://climateandsecurity.org/2013/03/admiral-locklear-climate-change-the-biggest-long-term-security-threat-in-the-pacific-region/.
14. "USPACOM's Admiral Locklear: Climate and Security in the Asia-Pacific," Center for Climate and Security, March 7, 2013, https://climateandsecurity.org/2013/03/uspacoms-admiral-locklear-climate-and-security-in-the-asia-pacific/.
15. Larry Liebert, "Inhofe Can't Budge an Admiral Who Says Climate Change Matters," Bloomberg, April 9, 2013, https://www.bloomberg.com/news/articles/2013-04-09/inhofe-can-t-budge-an-admiral-who-says-climate-change-matters.
16. Caitlin Werrell and Francesco Femia, "Climate Security and the U.S. Pacific Command Posture: Strategic Long-Term Challenges," Center for Climate and Security, April 20, 2015, https://climateandsecurity.org/2015/04/climate-security-and-the-u-s-pacific-command-posture-strategic-long-term-challenges/.
17. Iffat Idris, "Trends in Conflict and Stability in the Indo-Pacific," Institute of Development Studies, K4D Emerging Issues Report 42, 2020, https://doi.org/10.19088/K4D.2021.009.
18. Nitin Kumar, "Mumbai among Cities at Maximum Risk Due to Rising Sea

Levels: WMO Report," *Business Standard* (New Delhi), last updated February 15, 2023, https://www.business-standard.com/article/current-affairs/mumbai-among-cities-at-maximum-risk-due-to-rising-sea-levels-wmo-report-123021501213_1.html; M. Becker, M. Karpytchev, and A. Hu, "Increased Exposure of Coastal Cities to Sea-Level Rise Due to Internal Climate Variability," *Nature Climate Change* 13, no. 4 (April 2023): 367–74, https://doi.org/10.1038/s41558-023-01603-w; Anh Cao et al., "Future of Asian Deltaic Megacities under Sea Level Rise and Land Subsidence: Current Adaptation Pathways for Tokyo, Jakarta, Manila, and Ho Chi Minh City," *Current Opinion in Environmental Sustainability* 50 (June 1, 2021): 87–97, https://doi.org/10.1016/j.cosust.2021.02.010.

19. Michael Oppenheimer et al., "Sea Level Rise and Implications for Low-Lying Islands, Coasts and Communities," chap. 4 in *IPCC Special Report on the Ocean and Cryosphere in a Changing Climate*, edited by H.-O. Pörtner et al., 321–445 (Cambridge, UK: Cambridge University Press, 2019).
20. Erin Sikorsky, "China's Climate Security Vulnerabilities," Center for Climate and Security, Council on Strategic Risks, November 2022, https://councilonstrategicrisks.org/wp-content/uploads/2022/11/China-Climate-Security-Vulnerabilities-2022.pdf.
21. Joshua W. Busby, "Cyclones in South Asia: The Experiences of Myanmar, Bangladesh, and India," in *States and Nature: The Effects of Climate Change on Security*, 176–221 (Cambridge, UK: Cambridge University Press, 2022), 180.
22. Sarang Shidore, "Climate Security and Instability in the Bay of Bengal Region," Council on Foreign Relations, Center for Preventive Action, Discussion Paper Series on Managing Global Disorder no. 13, April 2023, https://www.cfr.org/report/climate-security-and-instability-bay-bengal-region.
23. Locklear, interview.
24. Mackenzie Allen, "Sherri Goodman on the Climatization of Security for the Bangladesh Institute of Peace and Security Studies," Center for Climate and Security, January 21, 2021, https://climateandsecurity.org/2021/01/sherri-goodman-on-the-climatization-of-security-for-the-bangladesh-institute-of-peace-and-security-studies/.
25. World Meteorological Organization, "World's Deadliest Tropical Cyclone Was 50 Years Ago," November 12, 2020, https://wmo.int/media/news/worlds-deadliest-tropical-cyclone-was-50-years-ago.
26. Munir Muniruzzaman and Shafqat Munir, interview by author, November 12, 2022.
27. Joshua W. Busby, *States and Nature: The Effects of Climate Change on Security* (Cambridge, UK: Cambridge University Press, 2022).
28. Bimal Kanti Paul, "Why Relatively Fewer People Died? The Case of

Bangladesh's Cyclone Sidr," *Natural Hazards* 50, no. 2 (2009): 289–304, https://doi.org/10.1007/s11069-008-9340-5.
29. Yazhou Sun, "Climate Migration Pushes Bangladesh's Megacity to the Brink," Bloomberg, June 28, 2022, https://www.bloomberg.com/news/features/2022-06-28/bangladesh-flooding-fuels-climate-migration-to-dhaka.
30. Natural Resources Defense Council, "Bangladesh: A Country Underwater, a Culture on the Move," September 13, 2018, https://www.nrdc.org/stories/bangladesh-country-underwater-culture-move.
31. Mubashar Hasan and Geoffrey Macdonald, "How Climate Change Deepens Bangladesh's Fragility," United States Institute of Peace, September 13, 2021, https://www.usip.org/publications/2021/09/how-climate-change-deepens-bangladeshs-fragility.
32. United Nations, Executive Office of the Secretary-General, "One Million Rohingya Refugees without Immediate Prospects for Return to Myanmar Five Years On, Secretary-General Says, Stressing Durable Solution to Crisis Is Critical," press release SG/SM/21424, August 24, 2022, https://press.un.org/en/2022/sgsm21424.doc.htm.
33. Human Rights Watch, "India: Investigate Alleged Border Force Killings," February 9, 2021, https://www.hrw.org/news/2021/02/09/india-investigate-alleged-border-force-killings.
34. Muniruzzaman and Munir, interview.
35. Royal Commission into National Natural Disaster Arrangements, "Royal Commission into National Natural Disaster Arrangements—Report," October 30, 2020, https://www.royalcommission.gov.au/natural-disasters/report.
36. Climate Council, "The Great Deluge: Australia's New Era of Unnatural Disasters," 2022, https://www.climatecouncil.org.au/resources/the-great-deluge-australias-new-era-of-unnatural-disasters/.
37. Elliot Parker, "Climate and Australia's National Security," The Forge, November 16, 2022, https://theforge.defence.gov.au/publications/climate-and-australias-national-security.
38. Kirsty Needham, "Australia Signs Security, Migration Pact with Pacific's Tuvalu," Reuters, November 10, 2023, https://www.reuters.com/world/asia-pacific/australia-offer-climate-refuge-all-residents-tuvalu-report-2023-11-10/.
39. Iberdrola, "Does Sea Level Rise Really Endanger Our Future?," accessed November 24, 2023, https://www.iberdrola.com/sustainability/sea-level-rise; John A. Church et al., "Sea Level Change," chap. 13 in Intergovernmental Panel on Climate Change, *Climate Change 2013: The Physical Science Basis; Contribution of Working Group I to the Fifth Assessment Report of the Intergovernmental*

Panel on Climate Change, edited by T. F. Stocker et al., 1137–1216 (Cambridge, UK: Cambridge University Press, 2013).

40. Mark Nevitt, "Climate Change and the Specter of Statelessness," *Georgetown Environmental Law Review* 35, no. 2 (2023): 331–57, Emory Legal Studies Research Paper no. 22-29, http://dx.doi.org/10.2139/ssrn.4223806, https://www.law.georgetown.edu/environmental-law-review/wp-content/uploads/sites/18/2024/01/Nevitt-Article.pdf.

41. Caitlin E. Werrell and Francesco Femia, eds., "The US Asia-Pacific Rebalance, National Security and Climate Change," Center for Climate and Security, November 2015, https://climateandsecurity.org/asiapacificrebalance/.

42. Woodrow Wilson International Center for Scholars, "Closing the Gap: Improving Early Warning for Climate Security Risks in the Pacific," November 14, 2019, https://www.wilsoncenter.org/event/closing-the-gap-improving-early-warning-for-climate-security-risks-the-pacific.

43. Woodrow Wilson International Center for Scholars, "Closing the Gap."

## Chapter 7: Imperiled Neighbors to the South

1. Elmer Roman, interview by author, June 16, 2023.
2. Bart Kenner et al., "Puerto Rico's Agricultural Economy in the Aftermath of Hurricanes Irma and Maria: A Brief Overview," US Department of Agriculture, Economic Research Service, Administrative Publication no. 114, April 2023, https://www.ers.usda.gov/webdocs/publications/106261/ap-114.pdf.
3. Max Zahn, "Puerto Rico's Power Grid Is Struggling 5 Years after Hurricane Maria. Here's Why," ABC News, September 22, 2022, https://abcnews.go.com/Technology/puerto-ricos-power-grid-struggling-years-hurricane-maria/story?id=90151141.
4. John D. Sutter, "130,000 Left Puerto Rico after Hurricane Maria, Census Bureau Says," CNN Health, December 19, 2018, https://www.cnn.com/2018/12/19/health/sutter-puerto-rico-census-update/index.html.
5. Roman, interview.
6. Caroline Kenny, "Trump Tosses Paper Towels into Puerto Rico Crowd," CNN Politics, October 3, 2017, https://www.cnn.com/2017/10/03/politics/donald-trump-paper-towels-puerto-rico/index.html; "Trump: Puerto Rico Not 'Real Catastrophe Like Katrina,'" BBC, October 3, 2017, https://www.bbc.com/news/av/world-us-canada-41492993.
7. Chuck Hagel, interview by author, June 27, 2023.
8. White House, "National Security Strategy," October 12, 2022, https://www.whitehouse.gov/wp-content/uploads/2022/10/Biden-Harris-Administrations-National-Security-Strategy-10.2022.pdf.

9. James Stavridis, interview by author, July 21, 2023.
10. Stavridis, interview.
11. World Wildlife Fund, "Climate Change Impacts in Latin America," accessed November 25, 2023, https://www.wwfca.org/en/our_work/climate_change_and_energy/climate_change_impacts_la/; Chantelle Burton et al., "South American Fires and Their Impacts on Ecosystems Increase with Continued Emissions," *Climate Resilience and Sustainability* 1 (2022): e8, https://doi.org/10.1002/cli2.8; World Meteorological Organization, "Climate Change Vicious Cycle Spirals in Latin America and Caribbean," July 5, 2023, https://wmo.int/news/media-centre/climate-change-vicious-cycle-spirals-latin-america-and-caribbean.
12. World Wildlife Fund, "Surveying Climate Change Impacts on Central America's Coral Reefs," March 19, 2007, https://www.wwfca.org/?96980/Surveying-climate-change-impacts-on-Central-Americas-coral-reefs.
13. Irene Mia and Juan Pablo Bickel, "How Climate Change Risks Further Destabilising Central America," International Institute for Strategic Studies, November 15, 2021, https://www.iiss.org/online-analysis/online-analysis//2021/11/how-climate-change-risks-further-destabilising-central-america.
14. Caitlin E. Werrell and Francesco Femia, eds., *Epicenters of Climate and Security: The New Geostrategic Landscape of the Anthropocene* (Washington, DC: Center for Climate and Security, June 2017), 94, https://climateandsecurity.org/wp-content/uploads/2017/06/epicenters-of-climate-and-security_the-new-geostrategic-landscape-of-the-anthropocene_2017_06_091.pdf.
15. Sam Meredith, "'We Are Not Drowning, We Are Fighting': Countries Vulnerable to Climate Disaster Issue Rallying Cry," CNBC, November 2, 2021, https://www.cnbc.com/2021/11/02/cop26-maldives-barbados-and-climate-activists-issue-warrior-cry-to-world.html.
16. "Statement of General Laura J. Richardson, Commander, United States Southern Command, before the 118th Congress House Armed Services Committee, March 8, 2023," https://armedservices.house.gov/hearings/full-committee-hearing-us-military-posture-and-national-security-challenges-north-and-south.
17. Christopher Klein, "Why the Construction of the Panama Canal Was So Difficult—and Deadly," History.com, October 25, 2021, October 25, 2021, updated September 15, 2023, https://www.history.com/news/panama-canal-construction-dangers.
18. History.com editors, "Panama Canal Turned Over to Panama," History.com, November 24, 2009, updated December 21, 2021, https://www.history.com/this-day-in-history/panama-canal-turned-over-to-panama.

19. My visit was made possible in part by Joe Reeder, then undersecretary of the army and the US representative to the Panama Canal Commission.
20. Wesley Clark, interview by author, April 25, 2023.
21. Stavridis, interview.
22. Laura Richardson, interview by author, June 30, 2023.
23. Richardson, interview.
24. Max Romero, interview by author, June 19, 2023.
25. Conor Finnegan, "Guatemala's President Calls for US to Help 'Build a Wall of Prosperity,'" ABC News, April 8, 2021, https://abcnews.go.com/Politics/biden-admin-surge-aid-central-america-offer-legal/story?id=76931611.
26. Andrew R. Arthur, "Report: Thousands of Unaccompanied Children in Border Patrol Custody," Center for Immigration Studies, January 6, 2022, https://cis.org/Arthur/Report-Thousands-Unaccompanied-Children-Border-Patrol-Custody.
27. Finnegan, "Guatemala's President Calls for US to Help 'Build a Wall of Prosperity.'"
28. Ellie Kaufman, "US Defense Secretary Tells US and NATO Troops in Bulgaria They Are Creating 'Trust' by Training Together," CNN, March 18, 2022, https://edition.cnn.com/europe/live-news/ukraine-russia-putin-news-03-18-22/h_e65a1ca12d3c55db3a1018d441fca3a9.
29. Romero, interview.
30. Richardson, interview.
31. "Coca Codo Sinclair Hydroelectric Project," Power Technology, October 1, 2020, https://www.power-technology.com/projects/coca-codo-sinclair-hydroelectric-project/?cf-view.
32. Nicholas Casey and Clifford Krauss, "It Doesn't Matter if Ecuador Can Afford This Dam. China Still Gets Paid," *New York Times*, December 24, 2018, https://www.nytimes.com/2018/12/24/world/americas/ecuador-china-dam.html.
33. Casey and Krauss, "It Doesn't Matter if Ecuador Can Afford This Dam."
34. Romero, interview.
35. Wesley Clark, interview by author, June 11, 2023.
36. Clark, interview.
37. Clark, interview.
38. Steve Brock et al., "The World Climate and Security Report 2020: A Product of the Expert Group of the International Military Council on Climate and Security," edited by Francesco Femia and Caitlin Werrell, February 2020, https://imccs.org/wp-content/uploads/2021/01/World-Climate-Security-Report-2020_2_13.pdf.

## Chapter 8: Climate Readiness on the Home Base

1. Ann Phillips, interview by author, July 21, 2023.
2. Phillips, interview.
3. Phillips, interview.
4. Phillips, interview.
5. Anthony Zinni, interview by author, August 2, 2022.
6. Phillips, interview.
7. Paul W. Valentine and Michael Weisskopf, "Aberdeen Civilian Officials Guilty in Toxic Waste Case," *Washington Post*, February 23, 1989, https://www.washingtonpost.com/archive/politics/1989/02/24/aberdeen-civilian-officials-guilty-in-toxic-waste-case/69b15b57-9279-4d56-ab95-4aa4f3da5cab/.
8. Phillips, interview.
9. Dennis McGinn, interview by author, December 2, 2022.
10. Leon Panetta, interview by author, August 1, 2023.
11. Panetta, interview.
12. Panetta, interview.
13. National Guard Association of the United States, "Deputy SECDEF: Military Impacted by Climate Change," September 26, 2023, https://www.ngaus.org/newsroom/deputy-secdef-military-impacted-climate-change; see also Quil Lawrence, "The National Guard Turns to Firefighting amid Worsening Climate Change," NPR, August 22, 2023, https://www.npr.org/2023/08/22/1195289223/the-national-guard-turns-to-firefighting-amid-worsening-climate-change.
14. Center for Climate and Security, Council on Strategic Risks, "Military Responses to Climate Hazards (MiRCH) Tracker," last updated November 1, 2023, accessed November 28, 2023, https://councilonstrategicrisks.org/ccs/mirch/.
15. Center for Climate and Security, Council on Strategic Risks, "Military Responses to Climate Hazards (MiRCH) Tracker."
16. Erich B. Smith, "Fighting Wildfires Almost Year-Round, Guard Preps for More," US Army, May 27, 2022, https://www.army.mil/article/257080/fighting_wildfires_almost_year_round_guard_preps_for_more.
17. Alice C. Hill and Tess Turner, "Burning Threats: How Wildfires Undermine U.S. National Security," Just Security, July 19, 2023, https://www.justsecurity.org/87265/burning-threats-how-wildfires-undermine-u-s-national-security/.
18. Charlsy Panzino, "Live-Fire Training Ignited Blaze at Fort Carson, Officials Say," *Army Times*, March 26, 2018, https://www.armytimes.com/news/your-army/2018/03/26/live-fire-training-ignited-blaze-at-fort-carson-officials-say/.
19. Marc Kodack, "Climate Change Driving Increase in Black Flag Days at 100 U.S. Military Installations," Center for Climate and Security, December

16, 2019, https://climateandsecurity.org/2019/12/climate-change-driving-increase-in-black-flag-days-at-100-u-s-military-installations/.
20. Meredith Berger, interview by author, August 2, 2023.
21. Berger, interview.
22. US Department of Defense, Office of the Undersecretary of Defense (Acquisition and Sustainment), "Department of Defense Draft Climate Adaptation Plan," report submitted to National Climate Task Force and federal chief sustainability officer, September 1, 2021, https://www.sustainability.gov/pdfs/dod-2021-cap.pdf.
23. Berger, interview.
24. Berger, interview.
25. "The Road to Resilience at Marine Corps Recruit Depot Parris Island," MCICOM Installation Energy Project, https://www.mcicom.marines.mil/Portals/57/Docs/GF-Energy/MCICOM-fact-sheet_Parris-Island-ESPC.pdf?ver=2019-10-11-103957-120.
26. Berger, interview.
27. US Department of Defense, Office of the Assistant Secretary of Defense (Energy, Installations & Environment), "2014 Climate Change Adaptation Roadmap," June 2014, https://www.acq.osd.mil/eie/downloads/CCARprint_wForward_e.pdf.
28. John Conger, interview by author, July 31, 2023.
29. Richard Kidd, interview by author, July 31, 2023.
30. Kidd, interview.
31. "Executive Order 13514 of October 5, 2009: Federal Leadership in Environmental, Energy, and Economic Performance," *Federal Register* 74, no. 194 (October 8, 2009), https://www.govinfo.gov/content/pkg/FR-2009-10-08/pdf/E9-24518.pdf; Federal Facilities Environmental Stewardship and Compliance Assistance Center, "EO 13514 (Archive)—Revoked by EO 13693 on March 19, 2015, Sec. 16(b)," March 19, 2015, https://www.fedcenter.gov/programs/eo13514/.
32. US Environmental Protection Agency, "Energy Independence and Security Act of 2007," last updated June 28, 2023, accessed November 28, 2023, https://www.epa.gov/greeningepa/energy-independence-and-security-act-2007.
33. Kidd, interview.
34. Kidd, interview.
35. Berger, interview.
36. Berger, interview.
37. Kidd, interview.
38. Jens Stoltenberg, "NATO Climate Change and Security Impact Assessment: The Secretary General's Report," 2nd ed., North Atlantic Treaty Organization, 2023,

https://www.nato.int/nato_static_fl2014/assets/pdf/2023/7/pdf/230711-climate-security-impact.pdf.

39. Virginia Department of Conservation and Recreation, Office of Governor Ralph S. Northam, "Virginia Coastal Resilience Master Plan, Phase 1," December 2021, https://www.dcr.virginia.gov/crmp/plan.
40. Commonwealth of Virginia, Office of the Governor, "Executive Order Number Twenty-Four: Increasing Virginia's Resilience to Sea Level Rise and Natural Hazards," November 2018, https://www.dcr.virginia.gov/crmp/document/Appendix-N-Executive-Order-24-Increasing-Virginia's-Resilience-to-Sea-Level-Rise-and-Natural-Hazards.pdf; Commonwealth of Virginia, Office of the Governor, "Executive Order Number Forty-Five: Floodplain Management Requirements and Planning Standards for State Agencies, Institutions, and Property," November 2019, https://www.dcr.virginia.gov/crmp/document/Appendix-O-Executive-Order-45-Floodplain-Management-Requirements-and-Planning-Standards-for-State-Agencies-Institutions-and-Property.pdf.
41. Phillips, interview.
42. Stoltenberg, "NATO Climate Change and Security Impact Assessment."
43. Panetta, interview.
44. Andrew Hoehn and Thom Shanker, *Age of Danger: Keeping America Safe in an Era of New Superpowers, New Weapons, and New Threats* (New York: Hachette Books, 2023).
45. Brady Dennis, "As Climate Change Worsens, Military Eyes Base of the Future on Gulf Coast," *Washington Post*, August 6, 2023, https://www.washingtonpost.com/climate-solutions/2023/08/06/climate-change-florida-military-tyndall/.
46. Dennis, "As Climate Change Worsens, Military Eyes Base of the Future."
47. Kidd, interview.
48. Mikaela Smith, "633d CES Taking Resiliency to a New Level," Joint Base Langley-Eustis, VA, July 31, 2023, https://www.jble.af.mil/News/Article-Display/Article/3477455/633d-ces-taking-resiliency-to-a-new-level/.
49. Kidd, interview.
50. Kidd, interview.
51. Panetta, interview.
52. Panetta, interview.
53. Dennis, "As Climate Change Worsens, Military Eyes Base of the Future."
54. Phillips, interview.
55. Panetta, interview.

## Chapter 9: Less Fuel, More Fight

1. Mike Henchen, interview by author, January 6, 2023.
2. Henchen, interview.

3. Sandra I. Erwin, "How Much Does the Pentagon Pay for a Gallon of Gas?," *National Defense*, January 4, 2010, https://www.nationaldefensemagazine.org/articles/2010/4/1/2010april-how-much-does-the-pentagon-pay-for-a-gallon-of-gas.
4. US Department of Defense, "Annual Energy Performance, Resilience, and Readiness Report: Fiscal Year 2022," June 8, 2023, https://www.acq.osd.mil/eie/Downloads/IE/FY22-AEPRR-Report.pdf.
5. CNA Military Advisory Board, "Powering America's Defense: Energy and the Risks to National Security," May 2009, https://www.cna.org/reports/2009/MAB_2-FINAL.pdf.
6. Greg Douquet, interview by author, August 11, 2023.
7. CNA Military Advisory Board, "Powering America's Defense."
8. Douquet, interview.
9. CNA Military Advisory Board, "Powering America's Defense."
10. Ray Mabus, interview by author, August 10, 2023.
11. CNA Military Advisory Board, "National Security and the Threat of Climate Change," May 1, 2007, https://www.cna.org/reports/2007/national-security-and-the-threat-of-climate-change.
12. Sherri Goodman, "Department of Defense Climate Change Programs," December 12, 1998, from author's files.
13. Dennis McGinn, interview by author, December 2, 2022.
14. US Department of Defense, Office of the Undersecretary of Defense for Acquisition and Technology, "Report of the Defense Science Board on More Capable Warfighting through Reduced Fuel Burden," May 2001, https://dsb.cto.mil/reports/2000s/ADA392666.pdf.
15. Amory B. Lovins et al., "Energy Efficiency Survey aboard USS *Princeton* CG-59," Rocky Mountain Institute, June 30, 2001, https://rmi.org/wp-content/uploads/2017/06/RMI_Energy_Efficiency_Survey_Aboard_2001.pdf.
16. US Department of Energy, "Energy for the War Fighter: The Department of Defense Operational Energy Strategy," June 14, 2011, https://www.energy.gov/articles/energy-war-fighter-department-defense-operational-energy-strategy.
17. US Department of Defense, Office of the Undersecretary of Defense for Acquisition, Technology, and Logistics, "Report of the Defense Science Board Task Force on DoD Energy Strategy, 'More Fight—Less Fuel,'" February 2008, https://apps.dtic.mil/sti/pdfs/ADA477619.pdf.
18. Mabus, interview.
19. Sharon Burke, interview by author, August 18, 2023.
20. Burke, interview.
21. Burke, interview.
22. Joe Bryan, interview by author, August 3, 2023.

23. US Department of Defense, Office of the Undersecretary of Defense (Acquisition and Sustainment), "Department of Defense Draft Climate Adaptation Plan," report submitted to National Climate Task Force and federal chief sustainability officer, September 1, 2021, https://www.sustainability.gov/pdfs/dod-2021-cap.pdf.
24. Frank Graff, "The Largest Floating Solar Farm in the Southeast Is in North Carolina," PBS North Carolina, June 30, 2023, https://www.pbsnc.org/blogs/science/the-largest-floating-solar-farm-is-in-the-southeast-is-in-north-carolina/.
25. David Roza, "Why the Military May Need Microgrids for Overseas Bases to Win a Near-Peer Fight," *Air and Space Forces Magazine*, October 31, 2023, https://www.airandspaceforces.com/air-force-military-overseas-bases-microgrids-power/.
26. Eielson Air Force Base, "Micro-reactor Pilot Program," https://www.eielson.af.mil/microreactor/.
27. Rachel Jacobson, interview by author, July 26, 2023.
28. Mabus, interview.
29. US Department of Defense, "Deputy Secretary of Defense Dr. Kathleen Hicks Remarks at Wayne State University, Detroit, Michigan, on Climate Change as a National Security Challenge," November 8, 2021, https://www.defense.gov/News/Transcripts/Transcript/Article/2838082/deputy-secretary-of-defense-dr-kathleen-hicks-remarks-at-wayne-state-university/.
30. Bryan, interview.
31. CNA Military Advisory Board, "Powering America's Economy: Energy Innovation at the Crossroads of National Security Challenges," July 2010, https://www.cna.org/reports/2010/powering-americas-economy-energy-innovation-at-the-crossroads-of-national-security-challenges.
32. Robert Hayward, interview by author, May 25, 2023.
33. Jim Garamone, "Hicks Defines Need to Focus DOD on Climate Change Threats," DOD News, August 30, 2023, https://www.defense.gov/News/News-Stories/Article/Article/3510772/hicks-defines-need-to-focus-dod-on-climate-change-threats/.
34. Henchen, interview.

## Chapter 10: Climate-Proofing Security

1. Center for Climate and Security, "Release: Mayors, Military Leaders and City Officials Raise Concerns about Sea Level Rise Threats to South Carolina's Military and Civilian Communities," August 7, 2018, https://climateandsecurity.org/2018/08/release-bipartisan-group-of-mayors-military-leaders-and-city-officials-raise-concerns-about-sea-level-rise-threats-to-south-carolinas-military-and-civilian-communities/.

2. US Department of Defense, "Statement by Secretary of Defense Lloyd J. Austin III on Tackling the Climate Crisis at Home and Abroad," January 27, 2021, https://www.defense.gov/News/Releases/Release/Article/2484504/statement-by-secretary-of-defense-lloyd-j-austin-iii-on-tackling-the-climate-cr/.
3. These are also the four pillars of the North Atlantic Treaty Organization's (NATO's) Climate Change and Security Action Plan, released in June 2021. See NATO, "NATO Climate Change and Security Action Plan," June 14, 2021, https://www.nato.int/cps/en/natohq/official_texts_185174.htm.
4. The acronym stands for Providing Research and End-user Products to Accelerate Readiness and Environmental Security. See US Department of Defense, "Enhancing Combat Capability—Mitigating Climate Risk: Department of Defense Budget, Fiscal Year (FY) 2024," March 2023, https://comptroller.defense.gov/Portals/45/Documents/defbudget/FY2024/PB_FY2024_ECC-Mitigating_Combat_Capability.pdf.
5. Joseph Clark, "Military Educators Converge on New Workshop Series to Discuss Resilience in Evolving Climate, Security Environment," US Department of Defense, news release, September 26, 2023, https://www.defense.gov/News/News-Stories/Article/Article/3539024/military-educators-converge-on-new-workshop-series-to-discuss-resilience-in-evo/.
6. Leon Panetta, interview by author, August 1, 2023.
7. David Titley, interview by author, August 10, 2022.
8. Nancy Walecki, "Tiny Climate Crises Are Adding Up to One Big Disaster," *Atlantic*, November 1, 2023, https://www.theatlantic.com/science/archive/2023/11/climate-disasters-low-intensity/675864/.
9. The term "climate-proofing" was coined by the Center for Climate and Security in 2017. See Caitlin E. Werrell et al., "A Responsibility to Prepare: Governing in an Age of Unprecedented Risk and Unprecedented Foresight," Center for Climate and Security, Briefer no. 38, August 7, 2017, https://climateandsecurity.org/wp-content/uploads/2017/12/a-responsibility-to-prepare_governing-in-an-age-of-unprecedented-risk-and-unprecedented-foresight_briefer-38.pdf.
10. Brady Dennis, "As Climate Change Worsens, Military Eyes Base of the Future on Gulf Coast," *Washington Post*, August 6, 2023, https://www.washingtonpost.com/climate-solutions/2023/08/06/climate-change-florida-military-tyndall/.
11. Cited in Jaron Z. Goldstein and Jason P. George, "Reducing Naval Fossil Fuel Consumption at Sea in the 21st Century," MBA Professional Project, Naval Postgraduate School, Monterey, CA, December 2021, https://apps.dtic.mil/sti/trecms/pdf/AD1164919.pdf.
12. Meredith Berger, interview by author, August 2, 2023.
13. US Marine Corps, "MCLB Albany First in DOD to Achieve Net Zero Energy

Milestone," May 25, 2022, https://www.marines.mil/News/News-Display/Article/3042811/mclb-albany-first-in-dod-to-achieve-net-zero-energy-milestone/.

14. National Intelligence Council, "National Intelligence Estimate: Climate Change and International Responses Increasing Challenges to US National Security through 2040," NIC-NIE-2021-10030-A, 2021, https://www.dni.gov/files/ODNI/documents/assessments/NIE_Climate_Change_and_National_Security.pdf.

15. "Statement by the Honorable Melissa Dalton, Assistant Secretary of Defense for Homeland Defense and Hemispheric Affairs, Office of the Secretary of Defense, before the 118th Congress Committee on Armed Services, U.S. House of Representatives, March 8, 2023," https://armedservices.house.gov/hearings/full-committee-hearing-us-military-posture-and-national-security-challenges-north-and-south.

16. US Department of the Navy, "SECNAV Delivers Remarks at Harvard Kennedy School," n.d., https://www.navy.mil/Press-Office/Speeches/display-speeches/Article/3538420/secnav-delivers-remarks-at-harvard-kennedy-school/.

17. See Center for Strategic and International Studies, Asia Maritime Transparency Initiative, "China Island Tracker," 2024, accessed March 18, 2024, https://amti.csis.org/island-tracker/china/.

18. US Department of Defense, "US Department of Defense Climate Assessment Tool," n.d., https://media.defense.gov/2021/Apr/05/2002614579/-1/-1/0/dod-climate-assessment-tool.pdf.

19. C. Todd Lopez, "DOD Makes Climate Assessment Tool Available to Partner Nations," US Department of Defense, April 21, 2023, https://www.defense.gov/News/News-Stories/Article/Article/3370578/dod-makes-climate-assessment-tool-available-to-partner-nations/.

20. Global Water Security Center, "SOUTHCOM, March–May 2024: Equator Divides Extreme Wet and Dry Conditions," February 26, 2024, https://ua-gwsc.org/wp-content/uploads/2024/02/QL20240226-NNSR-SOUTHCOM-El-Nino-watermarked.pdf.

21. Pauline Baudu and Sherri Goodman, "A Case for Increased Japan–US Cooperation on Climate Security in a Changing Strategic Context," Japan Up Close, May 10, 2023, https://japanupclose.web-japan.org/discussion/d20230510_1.html; Climate Council, "Quad Meeting First Step in Repairing Our Global Climate Reputation," May 23, 2022, https://www.climatecouncil.org.au/resources/quad-meeting-first-step-in-repairing-our-global-climate-reputation/.

22. Francesco Femia and Caitlin E. Werrell, "A Marshall Plan to Combat Climate

Change in the Asia-Pacific: The Missing Piece of the New U.S. Security Strategy," Center for Climate and Security, Briefer no. 8, February 7, 2012, https://climateandsecurity.org/wp-content/uploads/2012/04/a-marshall-plan-to-combat-climate-change-in-the-asia-pacific-the-missing-piece-of-the-new-u-s-security-strategy-_briefer-08.pdf.
23. Alan Rappeport, Lisa Friedman, and Keith Bradsher, "Yellen Urges China to Step Up Climate Finance Investments," *New York Times*, July 7, 2023, https://www.nytimes.com/2023/07/07/business/energy-environment/janet-yellen-beijing-climate.html.
24. Jim Garamone, "Austin, Papua New Guinea Leaders Discuss Plans for Defense Cooperation," DOD News, July 27, 2023, https://www.defense.gov/News/News-Stories/Article/Article/3472584/austin-papua-new-guinea-leaders-discuss-plans-for-defense-cooperation/.

# Index

Aberdeen Proving Ground, 37, 146
adaptation, importance of, 190–194
Afghanistan
 energy management and, 165–167, 173–177, 182
 military and environment in, 50, 89
 water scarcity and, 91–92
Africa, 100, 197
Africa Climate Summit 2023, 197
agriculture, impacts of climate change on, 125–126
Alaska, 147, 179
Albright, Madeleine, 48, 201
Allen, Thad, 76
alliances, importance of, 196–201
alternative energy sources, 171–172, 175, 177–179, 194–195
Amazon, 136
AMEC. *See* Arctic Military Environmental Cooperation
American Clean Energy and Security Act of 2009, 62

Anderson, Steve, 18–19
Arab Awakening, 93
Arctic Futures 2050, 74
Arctic Military Environmental Cooperation (AMEC), 79–84
Arctic region
 accidents in, 74–75
 cooperation and, 67–70, 78–84, 197
 Svalbard and, 71–73
 tipping points in climate and security and, 75–78
Arctic Report Card (NOAA), 70
Argentina, 120–121
Arizona, 149
artificial intelligence, 189–190
Asia. *See also specific countries*
 cooperation with leaders in, 106–107
 disaster alley and, 110–111
 extinction of countries in, 116–118
 identifying climate threats in region, 107–110
Assad, Bashar al-, 93

atmosphere, oceans and, 189
Atomic Energy Commission, 13
Austin, Lloyd J., 1, 187
Australia, 114–116, 199
automobile industry, 181
awareness, importance of, 188–190

Bagram Airfield, Afghanistan, 165–167
Bangladesh, 111–114
Barents Sea, 79
base closures, 27–31, 160, 185–186
bases of the future, 193
Baxter, Caroline, 190–191
Bay of Bengal, 110–111
Beckett, Margaret, 61
Berger, Meredith, 150–152, 154–155, 195
Bering Strait, 73
Bhola (Cyclone), 112
Biden, Joe, 62
Bikini Atoll, 107
biodiversity, 132–133
bipartisanship, 161–162
black flag days, 149
Black Start exercises, 152
blended fuels, 194
Blue Star Families, 144
Botswana, 197
Bowman, Frank "Skip," 53, 57, 71
Boxer, Barbara, 27
Bragg, Fort (North Carolina), 37–38
Brazil, 121, 136
Bromberg, Gideon, 96, 97
Browner, Carol, 49
Bryan, Joe, 178, 181–182
budget certification authority, 176
Bureau of Land Management, 39–40
Burke, Sharon, 63, 175–178
Busby, Joshua, 113
Bush, George H.W., 22–23, 50
Butts, Kent, 63
Byrd-Hagel Resolution, 44, 49

California
  climate-resilient infrastructure in, 151, 179
  costs of environmental violations in, 147
  endangered species and, 39–40, 143–144
  Moffett Field cleanup and, 35–36
California Air Resources Board, 147
Caribbean region. *See* Latin America and Caribbean region
Carns, Michael, 174
Carson, Fort, 149
Carter, Ash, 176
Carter, Jimmy, 128
Cavazos, Fort, 149
Center for Climate and Security, 5, 63, 105
Center for Naval Analyses (CNA), 50
Central Arctic Fisheries Agreement, 77
Charleston Naval Shipyard, 186
Chile, 120–121
China
  fishing in Arctic and, 77
  influence of in Indo-Pacific region, 106, 108, 115–116, 117, 197–198
  in Latin America and Caribbean region, 127, 132–134
  low-intensity conflicts with, 192
  mining in Africa and, 100
  oil supplies and, 174
  strategies for dealing with, 200
CINCPAC, 106
The Citadel, 185, 187–188, 192
Clark, Wesley, 128–129, 135–136
Clean Water Act, 146
Climate Adaptation Plan (DOD), 150–151
Climate Adaptation Roadmap (DOD; 2014), 152
Climate and Security Advisory Group, 5, 63

# INDEX

Climate Assessment Tool (CAT), 199. *See also* DOD Climate Assessment Tool
climate conflicts, 62
climate intelligence, 189
climate literacy, adaption and, 191–192
climate readiness.
    bird's-eye view of, 161–163
    climate-proofing of bases and forces for, 155–160
    historical approach to environmental issues and, 142–146
    increasing military awareness and, 146–150
    leadership and, 160–161
    as mission readiness, 150–155
    Superstorm Sandy and, 139–142
climate security as new field of study, 63
climate translators, 191–192
climate-informed decision-making, 150–151
Clinton, Bill, 22–23, 27–30, 44, 161–162
Clinton, Hillary, 61
closures of military bases, 27–31, 160, 185–186
CNA. *See* Center for Naval Analyses
CNA Military Advisory Board, overview of, 5, 52–54, 63
Coast Guard, 139–142
Coca Codo Sinclair hydroelectric facility, 134–135
coffee, 125–126
*Cole* (USS), 120
Colorado, 149
Conger, John, 152, 171
COP. *See* United Nations Conference of the Parties
Costa Rica, 121, 125
costs of energy, 172, 174, 175
Crimean invasion, 78

crops, 125–126
Cuban Missile Crisis, 11

Dabelko, Geoffrey, 63
DCAT. *See* DOD Climate Assessment Tool
de Leon, Rudy, 24
Defense Environmental Conference, 107
Defense Environmental International Cooperation (DEIC), 197
Defense Environmental Response Task Force, 34–35
Defense Nuclear Facilities Safety Board, 15
Defense Operational Resilience International Cooperation (DORIC), 197
Del Toro, Carlos, 195, 197–198
Delahunt, Bill, 32
DeMars, Bruce, 143
Department of Defense (DOD), 142, 147–148, 150–155, 167–171, 175–180. *See also* DOD Climate Assessment Tool
destabilization, 57
Deutch, John, 24
digital twins, 189–190
dignity, migration with, 117–118
disaster alley, 110–111
diseases, crops and, 125–126
disinformation, 77
distributed independent electrical grids, 178
DOD. *See* Department of Defense
DOD Climate Assessment Tool (DCAT), 160–161, 199
DORIC. *See* Defense Operational Resilience International Cooperation
Douquet, Greg, 169–170

drought, 50, 91–95, 125
drug cartels, 129

eagle, bald, 37
Earle, Naval Weapons Station, 152
Earth Summit (Rio de Janeiro), 22–23, 50
earthquakes, 121, 160
EcoPeace Middle East, 95–99, 197
Ecuador, 133–134
education, 98–99, 190–192
Eisenhower, Dwight D., 55–56
El Salvador, 125
electric vehicles, 180–181
Elson Air Force Base, 179
endangered species protection, 3, 36–40, 143–144
*Endurance* (HMS), 120
energy efficiency, 171–172, 174–175, 180–182, 194–195
energy independence, 183
Energy Independence and Security Act of 2007, 153
Energy Policy Act of 2005, 153
Engel, Rich, 63
Environmental Security Technology Certification Program (ESTCP), 193–194
Euphrates River, 94
Executive Order No. 13514, 153
exemptions, national security, 47–48
existential climate threats, 197–198
Expeditionary Strike Group 2 (Navy), 139–142

Farrell, Larry Jr., 171
Feinstein, Dianne, 27
Femia, Francesco, 63
firefighting chemicals, 18–19
fires. *See* wildfires
First Movers Coalition, 195

fishing, 77, 79, 88–89, 132–133, 200
*Fitzgerald* (USS), 120
Florida, 149, 158, 159, 193
fog of war, 55
Foggo, James, 67–68
forecasts, modeling and, 188–189
forever chemicals, 18, 194
fossil fuel supplies
    advanced technologies and, 180–182
    Bagram Airfield and, 165–167
    DOD energy consumption and, 167–171
    mission readiness and, 171–175
    progress in reducing reliance on, 182–184
    reducing reliance on, 175–180
Fritz, Oliver, 177
Fukushima Nuclear Disaster, 152
fungal diseases, 125–126

Ganges River, 110–111
Georgia, 159, 181
Gingrich, Newt, 44
Glenn, John, 14
Global Climate Change Security Oversight Act (2007), 62
Global Marshall Plan, 17
Goddard Institute for Space Studies, 54
Gore, Al, 17, 22–23, 44, 50
gray-zone tactics, 72–73
Great Green Fleet, 194
Green Blue Deal for Middle East, 96–98
Guatemala, 125, 132
Gunn, Lee, 103–104, 117
Guyana, 132

Hagel, Chuck, 124
Haiyan (Typhoon), 111
halon, 18–19
Hansen, James, 50, 54–55
Hayward, Robert, 85–86, 182–183

Heinze, April, 143
Henchen, Mike, 165–167, 182, 183–184
Hicks, Kathleen, 148, 180, 183
Hollings, Fritz, 186
Honduras, 125
Hurricane Sandy Rebuilding Task Force, 148
hurricanes. *See specific storms*
Hussein, Saddam, 169

ice thickness (Arctic), 67–68, 70–71
ICEX, 70
improvised explosive devices (IEDs), 173–174
India, 108, 114, 199
Indonesia, 109
Indo-Pacific region, cooperation and, 197
infrastructure, adaptation and, 190–191, 193
Inhofe, James, 64, 109, 162–163
International Military Council on Climate and Security, 5
Iran, 192
Iraq
    control of water supplies in, 94–95
    energy consumption and, 169–170
    energy instability as stressor in, 85–86, 182–183
    energy management and, 173–175
    military impacts on environment in, 89
Irwin, Fort (California), 39–40
Isabel (Hurricane), 159
ISIS, 94–95
Israel-Jordan Peace Treaty, 96–98

Jacobson, Rachel, 178–179, 181
Jakarta, Indonesia, 109
Janetos, Anthony, 54
Japan, 103, 152, 199
Jones, Anita, 193–194
Jordan, 96–98

Kelly Air Force Base (Texas), 35
Kennedy, Ted, 32–33
Kenya, 197
Kern, Paul, 53
Kerry, John, 32–33
Kidd, Richard, 153–154, 155, 158
King, Marcus, 92
Kiribati, 116–118
Kosmo, Jorgen, 79
Kratz, Kurt, 143
Krulak, Charles, 38
Kyoto Protocol, 43–49, 90

Langley-Eustis, Joint Base, 158–159
Latin America and Caribbean region. *See also specific countries*
    building resilience in, 131–132
    cooperation and, 126–131, 198
    democracy and, 132–135
    Elmer Roman and, 119–123
    environmental stewardship and, 135–137
    relationship with mainland U.S., 123–126
Lawrence Livermore National Laboratory, 17
Lejeune, Camp, 38
Liberty, Fort, 178–179
lidar mapping, 121
Lieberman-Warner Climate Security Act of 2008, 62
littoral combat ship (LCS) warships, 175
living shorelines, 158
Locklear, Samuel, 104–106, 108–112
locks, Panama Canal and, 128–129
Lopez, Joe, 26
Los Alamitos, California, 179
Los Alamos National Laboratory, 17
Lovins, Amory, 172–173
low-intensity conflicts/disasters, 192–193
Luke Air Force Base, 149

Mabus, Ray, 170–171, 174, 175, 180, 194
machine learning, importance of, 189–190
MAD. *See* mutual assured destruction
Mali, 92
María (Hurricane), 121–122
Marine Corps Logistics Base (Albany, GA), 195
marine expeditionary forces (MEF), 144
Marines, 151
Markey, Edward, 61
Marshall Islands, 107, 116–118
Maryland, 37, 181
Massachusetts Military Reservation (MMR), 32–34
Mattis, James, 173–174
*McCain, John S.* (USS), 120–122
McCurdy, David, 162
McGinn, Dennis, 146–147, 172–173
MEF. *See* marine expeditionary forces
Michael (Hurricane), 158
microgrids, 179
Middendorp, Tom, 91–92
Middle East. *See also specific countries*
   access to natural resources and stability in, 86–92
   changing priorities in, 99–101
   cooperation and, 197
   EcoPeace and, 95–99
   water, violence and, 92–95
migration
   Asia and, 114–118
   Australia and, 115–116
   global security and, 62
   Latin America and Caribbean region and, 124–126, 132
   Middle East and, 57–58, 93
   modeling of, 190
Military Responses to Climate Hazards (MiRCH) Tracker, 198

mining, illegal, 133
Misawa Air Base, 152
mission readiness, 171–175, 190–191, 192
mitigation, importance of, 194–195
MMR. *See* Massachusetts Military Reservation
modeling, 188–189, 199
Moffett Field Naval Air Station (California), 35–36
"More Capable Warfighting through Reduced Fuel Burden" study, 49, 172
"More Fight- Less Fuel" report, 174
Morehouse, Tom, 18–19, 143, 174, 176
Mosul Dam, 94
Mottley, Mia, 126
Munir, Shafqat, 112
Muniruzzaman, Munir, 112–113
Murkowski, Lisa, 73
Murmansk, 79
Murray, Robert J., 9, 51
mutual assured destruction (MAD), 11
mutually assured resilience, 100–101
Myanmar, 114

nation extinction, 116–118
National Defense Authorization Act, 10–11
National Guard, 148
National Intelligence Estimate on Climate Change, 73
NATO. *See* North Atlantic Treaty Organization
NATO Allied Command Transformation, 156
natural gas, Middle East and, 87–88
natural science, social science and, 190
The Nature Conservancy, 158
Near (James B.) Center for Climate Studies, 185, 187–188, 192
near-Arctic states, 77

net-zero energy, 194–195
neutron bomb, 8–9
New Jersey, 139–142, 147–148, 152, 157–158
New York, 12, 139–142, 147–148, 157–158
*New York Times Magazine*, 62
Newell, Peter, 176
Nicaragua, 125
Norfolk, Virginia, 151, 155–159
Normandy invasion, 55–56
North Atlantic Treaty Organization (NATO), 47, 196
North Carolina, 178–179
North Korea, 109
Northam, Ralph, 156
Northern Sea Route, 75–76
Norway, 71–73, 79
nuclear fuel, 78–84
Nuclear Non-Proliferation Act of 1978, 14
nuclear power plants, safety oversight at, 12–16
Nuclear Regulatory Commission, 15
Nunn, Sam, 7, 9–11, 13–14, 16–17, 27

Obama, Barack, 148, 153, 174
oceans, atmosphere and, 189
Office of Naval Research, 120
Oglala Sioux Tribe, 143–144
Ohio, 144
oil supplies, 75–76, 87–88
Oklahoma, 162–163
opportunity multiplier
  adaptation and, 190–194
  alliances and, 196–201
  awareness and, 188–190
  creation of, 132, 137
  mitigation and, 194–195
Oslo Accords, 96
Owens, William, 33

Pakistan, 112
Panama, 127–129
Panama Canal, 127–128, 135
Panetta, Leon, 147–148, 157, 160, 161, 191
Papua New Guinea, 200
Parris Island, Marine Corps Recruit Depot, 151, 152
Parthemore, Christine, 63
Pendleton, Marine Corps Base, 143–144
Perry, William, 79, 185–186
PFAS. *See* polyfluoroalkyl substances
Phillips, Ann, 139–142, 144–145, 156–157, 161
plutonium, 15
Polar Silk Road, 75
polyfluoroalkyl substances (PFAS), 18, 194
PREPARES program, 190
*Princeton* (USS), 173
private sector, partnerships with, 180–182
Prueher, Joseph, 52, 61–62, 106–107, 107–108
Puerto Rico, 119–123, 131
Pulwarty, Roger, 192
Punaro, Arnold, 7–9, 16
Putin, Vladimir, 75–76, 84, 196

Quadrilateral Security Dialogue (QUAD), 199
quantum technology, importance of, 189–190

RABs. *See* Restoration Advisory Boards
range cleanup, 143–144
*Ranger* (USS), 147
Rapid Equipping Force, 176
resilience, 100–101, 150–155
Restoration Advisory Boards (RABs), 35–36
Rice, Rick, 143

Richardson, Laura, 126, 131, 132–133, 137
Riley, Fort (Kansas), 37
Rocky Flats plutonium processing plant, 11–12, 15
Rocky Mountain Institute, 184
Rohingya refugees, 114
Roman, Elmer, 119–123
Romero, Max, 131, 132–135
Roosevelt, Theodore, 127
Rosatom, 76
Ross, Rachel, 177
Roughead, Gary, 70
Russell, Richard, 10
Russia
  Arctic and, 70, 72–73, 75–76, 78
  collaborating with, 78–84
  oil supplies and, 174
  strategic competition with, 192, 196, 200
  targeting of energy infrastructure by, 179
Ryan, Steve, 14

salvage divers, 119–120
San Diego, California, 151
Sandia National Laboratory, 17
Sandy (Hurricane/Superstorm), 139–142, 144, 147–148, 152, 157–158
Sandy Hook, Coast Guard Station, 139–142
Saudi Arabia, 170–171
Schlesinger, James, 174
Sea Angel, Operation, 112–113
sea level rise, statistics on, 116
Select Committee on Energy Independence and Global Warming, 61
Senate Armed Services Committee, 14, 15, 17, 61
Senate Governmental Affairs Committee, 14

September 11, 2001 terrorist attacks, 173, 192
SERDP. *See* Strategic Environmental Research and Development Program
shipping routes, 75
Siegel, Lenny, 35–36
Sikes, Joe, 143
Sill, Fort, 162–163
Simmons, P. J., 63
Slocombe, Walter, 8
smart power systems, 178
Smith, Larry, 24
Smithsonian Tropical Research Institute (Panama), 128
social science, natural science and, 190
Soderberg, Nancy, 29
soft power strategies, 104, 110–111
solar power, 177–179, 182
Somalia, 88–89
South Carolina, 185–188, 192
South Dakota, 144–145
Stagg, James, 55–56
Stavridis, James, 124, 129
Stevens (Ted) Center for Arctic Security Studies, 197
Stewart, Fort, 159
Strategic Environmental Research and Development Program (SERDP), 17–18, 194
Sullivan, Gordon R., 36–40, 52, 58, 59
Superfund sites, 15, 24–25
sustainable development, 136
Svalbard archipelago, 71–73
Syria, 92–93
systems thinking, improving, 189

Taleb, Yana Abu, 97
Taliban, 91
technology adaptation, 193
territories of United States, 119–123, 131

tether of fuel, 173–174
Texas, 149
threat multiplier, 4, 60, 62, 91–92, 137, 188
Three Mile Island accident, 12
Tigris River, 94
Tinker Air Force Base, 163
Titley, David, 70–71, 191
Tong, Anote, 117–118
Topshee, Angus, 76–77
tortoise, desert, 39
Truly, Dick, 49, 53, 58, 61–62, 172, 174
Trump, Donald, 122
Tuvalu, 115–118
Tyndall Air Force Base, 158, 193

Ukraine, 73, 75, 179, 192, 196
United Nations Climate Change Conference (COP), 98, 126, 196
United Nations Framework Convention on Climate Change (UN-FCCC), 22
United States Military Academy West Point, 181, 192
US AFRICOM, 99–100
US Fourth Fleet, 129
US Global Water Security Center, 199
US Indo-Pacific Command, 106
US SOUTHCOM, 124, 126, 128–131, 136–137

Vanuatu, 200
Veeravalli, Swathi, 99–101
Venezuela, 174
Vinson, Carl, 10
Virginia, 151, 158–159

Wald, Chuck, 47, 61–62, 90–91
Walker, Tracey, 143
war games, 74
Warner, John, 61, 64
Wasserman, Lee, 52
water scarcity, 50, 91–92
weaponization of water, 93–94
Weiss, Leonard, 14
Werrell, Caitlin, 63
West Point, 181, 192
White, Kate, 160
wildfires, 115–116, 148–150
wildlife refuges, 36
wind turbines, 182
Wirth, Tim, 136–137
woodpecker, red-cockaded, 37–38
Wormuth, Christine, 179
Wright-Patterson Air Force Base, 144

Yellen, Janet, 200

Zinni, Anthony, 52, 57, 88–89, 143–144

## About the Author

Photo credit: Kristina Sherk

Sherri Goodman has been a leader in environmental, energy, and climate security since she served as the first deputy undersecretary of defense (environmental security). Today, she is secretary general of the International Military Council on Climate and Security and a senior fellow at the Woodrow Wilson International Center for Scholars. She is also the founder and former executive director of the CNA Military Advisory Board, and she chairs the board of the Council on Strategic Risks, which includes the Center for Climate and Security.

She is credited with educating a generation of US military and government officials about the nexus between climate change and national security, using her famous coinage, "threat multiplier," to fundamentally reshape this field.

Sherri has twice received the Department of Defense Distinguished Service Award. She also received the Gold Medal Award from the National Defense Industrial Organization and the US Environmental Protection Agency's Climate Change Award. She received an honorary doctorate from Amherst College and the Lifetime Achievement Award from the Environmental Peacebuilding Association.

She has testified before numerous congressional committees and has shared her voice through outlets such as CNN, the BBC, NPR, CBS, the *New York Times*, the *Washington Post*, *Politico*, *Foreign Policy*, and *NATO Review*, among many others. She has taught classes and given lectures at numerous universities and has appeared in several films about climate change, including *The Age of Consequences*, *Journey to Planet Earth*, *Carbon Nation*, and *From Paris to Pittsburgh*.

Sherri holds degrees from Harvard Law School, Harvard Kennedy School, and Amherst College.

# About Island Press

Since 1984, the nonprofit organization Island Press has been stimulating, shaping, and communicating ideas that are essential for solving environmental problems worldwide. With more than 1,000 titles in print and some 30 new releases each year, we are the nation's leading publisher on environmental issues. We identify innovative thinkers and emerging trends in the environmental field. We work with world-renowned experts and authors to develop cross-disciplinary solutions to environmental challenges.

Island Press designs and executes educational campaigns, in conjunction with our authors, to communicate their critical messages in print, in person, and online using the latest technologies, innovative programs, and the media. Our goal is to reach targeted audiences—scientists, policy makers, environmental advocates, urban planners, the media, and concerned citizens—with information that can be used to create the framework for long-term ecological health and human well-being.

Island Press gratefully acknowledges major support from The Bobolink Foundation, Caldera Foundation, The Curtis and Edith Munson Foundation, The Forrest C. and Frances H. Lattner Foundation, The JPB Foundation, The Kresge Foundation, The Summit Charitable Foundation, Inc., and many other generous organizations and individuals.

The opinions expressed in this book are those of the author(s) and do not necessarily reflect the views of our supporters.